Data Analysis and Presentation Skills

Data Analysis and Presentation Skills

An Introduction for the Life and Medical Sciences

Jackie Willis
Coventry University, UK

John Wiley & Sons, Ltd

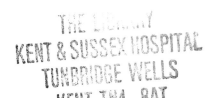
Copyright © 2004   John Wiley & Sons Ltd, The Atrium, Southern Gate, Chichester,
West Sussex PO19 8SQ, England

Telephone (+44) 1243 779777

E-mail (for orders and customer service enquiries): cs-books@wiley.co.uk
Visit our Home Page on www.wileyeurope.com or www.wiley.com

Other Wiley Editorial Offices

John Wiley & Sons Inc., 111 River Street, Hoboken, NJ 07030, USA

Jossey-Bass, 989 Market Street, San Francisco, CA 94103-1741, USA

Wiley-VCH Verlag GmbH, Boschstr. 12, D-69469 Weinheim, Germany

John Wiley & Sons Australia Ltd, 33 Park Road, Milton, Queensland 4064, Australia

John Wiley & Sons (Asia) Pte Ltd, 2 Clementi Loop #02-01, Jin Xing Distripark, Singapore
129809

John Wiley & Sons Canada Ltd, 22 Worcester Road, Etobicoke, Ontario, Canada M9W 1L1

Wiley also publishes its books in a variety of electronic formats. Some content that appears in
print may not be available in electronic books.

Library of Congress Cataloging-in-Publication Data

Willis, Jackie.
    Data analysis and presentation skills : an introduction for the life and
    medical sciences / Jackie Willis.
    p. cm.
    Includes bibliographical references and index.
    ISBN 0-470-85273-9 (acid-free paper) – ISBN 0-470-85274-7 (pbk. : acid-free paper)
    1. Research – Data processing.   2. Research – Technique.
    3. Medicine – Research – Data processing.   I. Title.
    Q180.55.E4W45 2004
    001.4'2'0285–dc22   2004015412

British Library Cataloguing in Publication Data

A catalogue record for this book is available from the British Library

ISBN 0 470 85273 9 hardback
     0 470 85274 7 paperback

Typeset in $10\frac{1}{2}$ /$13\frac{1}{2}$ pt Sabon by Dobbie Typesetting Ltd, Tavistock, Devon
Printed and bound in Great Britain by TJ International, Padstow, Cornwall
This book is printed on acid-free paper responsibly manufactured from sustainable forestry
in which at least two trees are planted for each one used for paper production.

To Malcolm, James, Jennifer and my parents

Contents

Preface

Science is a discipline that involves the planning of experiments, collecting data and evaluating the results. As an undergraduate, skills need to be developed in researching information, designing experiments then analysing and presenting the data produced. This book provides an introduction to data analysis and the techniques that may be used in presenting information for dissemination to the scientific community. These are demonstrated by using the standard packages available as part of Microsoft Professional/Microsoft Office, Excel and PowerPoint, an Internet browser and e-mail client software. Using Excel you will learn how to perform calculations on spreadsheets, present charts and graphs and perform statistical analyses. Applying PowerPoint you will be shown how to prepare information in the form of a poster or as slides for a seminar.

The mere mention of statistics is usually enough to provoke some disquiet amongst most students, frequently because of its association with mathematics. It is unfortunate that many undergraduates are unable to integrate the applications of statistics into their studies and feel at a loss to know what test to use or how to interpret the results of analyses. The approach taken in this book is to show how statistics may be applied and, by using the easily accessible functions in Excel, perform statistical analyses and explain the results that have been obtained. This is a tactic that has proved successful over the past several years and, as graduates, students have commented on how useful they found the approach, subsequently feeling more confident at moving on to use the more dedicated statistical packages such as MiniTab or SPSS.

Data Analysis and Presentation Skills by Jackie Willis.
© 2004 John Wiley & Sons, Ltd   ISBN 0470852739 (cased)   ISBN 0470852747 (paperback)

This book does not intend to enter into any in-depth discussions about the theoretical aspects of statistics. There are many excellent textbooks already available that deal with statistical concepts and provide full details of how tests are applied to the medical and life sciences. It is hoped that anyone who works through this book will then supplement their knowledge by consulting more demanding texts to build up their understanding of this essential subject.

What is important is that students lose their reluctance to perform data analysis by gaining confidence in how to use and apply the standard packages, so widely available. This book takes a very 'hands on' approach and each section will take the reader through several processes explaining each step in detail. Hints and tips are provided on presentation skills, use of statistics and researching using the Internet. The first section works through the basics of working in the Windows environment, the standard platform from which most software packages are launched. The reader should therefore be able to progress through the book whether they are already an experienced user or a complete novice who has had very little contact with computers before.

Complete support for the material contained in the book is to be found at the Wiley website: www.wileyeurope.com/go/WillisData. Here the reader will be able to look at worked examples of problems, view PowerPoint presentations and find lists of useful links to other sites. The website will be regularly updated so there should always be something new to find to reinforce the material in the book.

Acknowledgements

I would like to thank colleagues at Coventry University for their support, particularly Professor Phil Harris without whose encouragement this book would certainly never have been attempted in the first place. I am also grateful for the steadfast support of my family and to my daughter, Jennifer, for performing the painstaking proofreading of the final manuscript. Finally I would like to thank you, the reader, for purchasing this book. I hope that you will find it a useful aid to your studies and that it will play its part in developing a long and successful career in whatever branch of science you are studying.

Jackie Willis

1

Working in the Windows Environment

Microsoft Windows is the standard operating system from which all Microsoft applications are accessed. If you are unfamiliar with this package then it is essential that you work your way through this section before moving on to the remainder of the book. If you are already familiar with how Windows operates you may just want to read through this section to make sure that you can perform all of the functions described and then move on to Section 2.

1.1 Basic computing terms

Every computer system consists of a series of components, these are known as the hardware. Typically a computer system contains a Central Processing Unit (CPU) which includes the main microchip responsible for running the computer, housed in a tower system that also accommodates the hard drive. The hard drive is a storage device that is used to store the programs that will run the computer. These programs are the instructions that enable the computer to perform specific tasks. Each type of dedicated program is known as the computer software. The information that is exchanged between the computer and its human user may be viewed on a monitor. A keyboard and a mouse are used for the user to input information that the computer then processes using the software provided. Information may be stored on the hard drive or on a floppy disk or writable CD using an appropriate drive on the

Data Analysis and Presentation Skills by Jackie Willis.
© 2004 John Wiley & Sons, Ltd   ISBN 0470852739 (cased)   ISBN 0470852747 (paperback)

computer. Information may be printed and is known as a hard copy. This may vary from written material to photographs and other images.

The majority of computer systems use an operating system known as Windows, provided by the Microsoft Corporation. This has become standard software used as a platform from which other programs may be accessed and utilized. If you are using a computer on a university network then you are probably using Windows NT. If you are running your computer with the software accessed from the hard drive then you may be using Windows 2000, Windows XP or an even earlier version, depending on how old your system is. Whatever the version, the way in which the software functions is the same; it is just the appearance of the information on the screen or the way files are accessed which tends to differ slightly.

1.2 Working in Windows

All of the work you do is contained within a rectangular area of the screen known as the *window*. The background on which the windows are placed is the *desktop*. Each *application* that you work with through Windows (such as the word processing package Word and the spreadsheet application Excel) are represented by small graphical symbols known as *icons*. Your actions in Windows are carried out by using either the mouse or the keyboard, depending on the task in hand.

Using the mouse

The mouse is a mobile device that may be completely wireless or attached to the back of the computer by a lead. The mouse usually has two large buttons (left and right) and sometimes a middle wheel. Different functions may be selected by clicking the left or right buttons, whereas the wheel is used as a means to scroll up and down the screen quickly.

The most frequently used button on the mouse is the left button. This is used to move a pointer across the screen to select from menus containing various options, or to display *pull down* menus containing more options. The main actions of the mouse are to click and drag and this mainly applies to the left mouse button. The actions of the mouse are summarized in Table 1.1.

The keyboard

The computer keyboard is the same as you would find on a typewriter (QWERTY) for the general layout of the letters and numbers, but there are

Table 1.1 Actions for the left mouse button

Action	Mechanism	Uses
Click	Push down and release the left-hand mouse button	1. Selects an option from an on-screen menu 2. Repositions the insertion point in a document 3. Selects an option on the screen such as a program icon
Drag	Hold down the left-hand mouse button whilst moving (dragging) the mouse pointer across the screen	1. Highlights characters/blocks of text on the screen 2. Moves an object across the screen
Double click	Click the left-hand mouse button twice, rapidly	Selects and runs an application

some other important keys that may have special functions in an application. The numeric keys are also repeated on the right-hand side of the keyboard and are set out in blocks of three; these may be used when the NUM LOCK key is on (press the NUM LOCK key once and a light should indicate that is it on. Press again to switch the NUM LOCK off). Pressing down the CAPS LOCK key will result in all of the letters typed appearing in upper case. Pressing the key again will turn the CAPS LOCK off.

There is also a set of keys at the top of the keyboard labelled F1 to F12. These are known as the function keys as some applications will perform specific operations when one of these are pressed. These keys will not be described for any of the applications you will be using as all the functions can be carried out using the mouse.

The cursor

When using a program in Windows, the insertion point in a document or spreadsheet is shown by a flashing black line known as the cursor. When using the mouse a black line (⎀) or arrow may be moved across the screen to enable the user to reposition the cursor; by clicking with the left mouse button the cursor is moved to the point indicated.

There are some other important keys that you need to become familiar with. These are listed in Table 1.2.

Table 1.2   Important keys and their function

Key symbol	Name	Function
$\hookleftarrow$	Enter (or Return)	1. Enters a command or information. Returns the insertion to the left-hand side of the screen, one line down in a word processing package or into the next cell in a spreadsheet
$\leftarrow$	Backspace	Moves the cursor back by one character
$\mid\leftarrow$ $\rightarrow\mid$	Tab	Moves the cursor along to pre-set (or user-defined) positions (tab stops) on the screen
CAPS LOCK	Capitals lock	One press of the key will cause all letters to be typed in upper case (switched on). Press again to switch off
$\uparrow$	Shift	1. Causes letter characters to be typed in upper case (when the Caps Lock is off) 2. Produces symbols shown over number keys and other characters
Alt	Alternative	Key used in conjunction with other character keys in an application to bring about a specific function
Ctrl	Control	As for the Alt key
$\uparrow$ $\leftarrow \rightarrow$ $\downarrow$	Arrow keys	These move the cursor up, down, left or right in a document or spreadsheet or move the selection in a pull down menu

Opening an application

When you first start Windows, the screen opens with all of the applications (programs) appearing as icons. To enter the application you simply double click the left button on the mouse over the icon. The icon changes colour to indicate that the application is loading. Another way to open an application is by clicking the Start button at the bottom left-hand side of the screen. A menu of options opens. By going to Programs the list of applications appears from which you can select the one that you want. Using either method, open Microsoft Word.

Closing an application

To close and exit a window, click on the 'X' button at the top right of the application window. (N.B. there are usually two 'X' buttons, one for the specific file that you are working on, the other for the application). Try closing Microsoft Word. This operation is also accomplished by clicking on File and then choosing Exit from the drop down menu.

Minimizing

Sometimes you may want to put aside an application without having to close it whilst you use a different program. Open Excel by double clicking on the icon. The window may now be minimized without closing the application by clicking on the '–' button at the top right-hand corner of the application. When the application is minimized you should see the desktop once more. At the bottom left (on the status bar) you will see a button labelled Excel-Book 1 (or the name of the file that you were working on). Clicking on the button re-opens the program.

The Restore button, in between the minimize and close buttons, will control how much of the screen the window will take (it appears as two overlapping squares). By clicking on this button you should see the window reduce in size and the desktop appear behind. If you are working on two files in two applications, making the window smaller in this way enables you to switch between each application easily whilst being able to see what is happening in each window at the same time.

Note the change in appearance of the Restore button. As the window reduces, only one square is seen on the button. Click on the same button to maximize the window once more (note the change in appearance of the button to its original form).

Re-sizing

Application windows can also be re-sized. Move your mouse pointer until it is on one of the borders of the window. The pointer will change to a double arrow that indicates the direction you can re-size the window. Drag the border to the size that you require. A vertical and horizontal border can be re-sized at the same time by taking the mouse pointer to the corner of the window until a diagonal double arrow appears. Now drag the borders to the required size.

Moving a window

To move a window to a different place on the desktop, click on the Title bar and drag it to a new position. Try moving the Excel application so that it appears in the centre of the screen.

It is worthwhile spending some time to familiarize yourself with the different applications available from the desktop. Try the minimize, maximize and restore features until you feel confident in using them.

1.3 General functions in Microsoft applications

Within all of the Microsoft applications there are a common set of buttons which may be used as short cuts to perform various operations. Before moving on it is worthwhile going into Microsoft Word to practise using some of these easily accessible functions. From the desktop open Microsoft Word by either using the Start option or clicking the icon on the desktop as outlined in section 1.2.

At the top of the screen you will see various options (File, Edit, View, etc.) as seen in Figure 1.1. All of these options have drop down menus for a variety of features that can be used in Word or other Microsoft Office programs. To make document processing easier and faster there are also *toolbars* (which can be customized to your own preferences): the Standard toolbar and the Formatting toolbar. A summary of the commands available from these toolbars is given below.

Figure 1.1   Toolbars in Microsoft applications

Commands from the Standard toolbar

The Standard toolbar has buttons that are common to all Microsoft Office programs. Use the mouse to point at each button on the toolbar and a yellow 'flag' will appear that describes the function of each tool.

Figure 1.2   New, Open and Save buttons

New – create a new document (Word); new workbook (Excel); new presentation (PowerPoint).

Open – open a file saved on disk. A window will appear in which to search for the document from its location on the drive, e.g. hard disk (C:/) or floppy disk (A:/).

Save – save the current document you are working on. You will be prompted to select to which location you want to save the work and provide a suitable filename. The suffix to the filename indicates the package in which the file has been created. For Word this is '.doc', for Excel '.xls' and '.ppt' for PowerPoint.

Figure 1.3   Print, Print Preview, Spelling and Grammar buttons

Print – prints your current document providing a printer is attached or you have access to a printer through a network.

Print Preview – Allows preview of one page or several pages to see how the text fits into the page format. A more detailed description of this feature appears in Section 3.

Spelling and Grammar – allows spellchecking of all or a selection of your document. The software will automatically search the document for spelling mistakes as you are typing (and will correct some automatically). Any that it cannot match with its resident dictionary will be shown in the spellchecker window. You may then choose to change or ignore the highlighted word. The resident dictionary will not contain specialized words such as 'eukaryotic', so you will need to check these spellings for yourself, together with any names that are not recognized. You should also proofread your document to check for any typing mistakes that cause a word to be misused, e.g. *form* in place of *from*. It should become routine to spellcheck and proofread every document before printing.

Figure 1.4   Cut, Copy, Paste and Format Painter buttons

Cut – items can be highlighted and removed from the document completely or may simply be taken to a different place in the document and *Pasted* into a new position. This is a very useful feature for editing your work and shows the versatility of preparing drafts using a word processing package as opposed to pen and paper. The Cut command may also be used to cut drawn objects, photos or clip art images that are created in a document.

Copy – text or objects can be highlighted and copied elsewhere in the document. Using Copy will not delete the highlighted item as in the Cut command. Items that are cut or copied are said to be held on a *clipboard*. Items are then pasted back into the document from the clipboard.

Paste – Paste is used to insert text or another object at the required place within the document or may be used to move items from one document to another (sometimes from one application to another).

Format Painter – this is a feature that allows you to copy the format from one part of a document to another, or it may be applied from one document to another. It is used by firstly selecting the text or item whose format you would like to be copied elsewhere in the document. Then click the Format Painter button and highlight the items whose format is to be changed.

Figure 1.5   Undo and Redo buttons

Undo and **Redo** – Undo will remove the last change that you made to a document and Redo will change it back again. Applications allow multiple undo's and redo's to many tasks that have been performed, but some functions cannot be undone. This can be a powerful editing feature when you are finalizing a document.

Figure 1.6   Drawing button

Drawing – the drawing feature allows you to insert images into your document or to draw your own pictures and place them in your document.

Figure 1.7   Zoom control

Zoom Control – this zooms in and out of your document (and is sometimes a more helpful feature than Print Preview). Setting the control to 50% will show a full A4 sheet on your screen. Editing may be performed whilst zoomed in or out.

Figure 1.8   Help button

Help – help is available for all topics. By clicking on the Help button and typing in keywords associated with a function that you do not understand a search is done and a list of items containing the keyword is shown. You are then able to select the topic that is the most appropriate to your query.

The Formatting toolbar

This allows you to format your document as you are looking at it on screen as opposed to selecting items from the Format menu. These features appear in all Microsoft Office applications. Formatting tools may be seen in Figure 1.9 and a brief description of their functions is given below.

Figure 1.9   Formatting toolbar

Change of font (typesetting); the default is usually Times New Roman
Change of size of the text (point size); the default is usually 10 point
Emboldening text (B)
Changing standard text to italics (*I*)
Underlining (<u>U</u>)
Aligning text – to the left
           – in the centre (useful for centring titles)
           – to the right
Text justification (to even out spaces between words)
Line numbering
Insertion of bullet points
Increasing and decreasing indents to paragraphs
Introduction of text boxes to documents

Inserting additional toolbars

Sometimes it is useful to insert additional toolbars other than the Standard and Formatting toolbars that are shown by default as an application is opened.

If we are likely to be incorporating pictures into a document then having access to the Picture toolbar may be useful. To accomplish this, click on View and from the list of options obtained by scrolling down to Toolbars, select Picture. The Picture toolbar should now be displayed above the Standard toolbar. This allows you to insert pictures from file or the web and then format them.

This introductory section has outlined the features of the Windows environment. We will now move on to employ these functions as we learn how to use and apply programs in Microsoft Office.

2

Researching and Planning Scientific Investigations

The very nature of science is to pose questions and seek answers. In order to widen the boundaries of our knowledge we have to design experiments that will produce the answers we are looking for. The planning stage is crucial to the success of the investigation. Without carefully considering what information is available to us before we begin, time and resources can be wasted or the experiments we design will be flawed. In this section we will look at ways in which information can be accessed and evaluated using Internet resources and then consider some of the issues in experimental design.

2.1 Sources of information

If we are going to plan a scientific investigation, before considering what laboratory work or trials to perform, we need to thoroughly research the background information to our studies. This involves consulting information from a variety of sources including books, the media and scientific papers to find out what the current thinking is in our field of interest. One of the most frequently used resources to access information, other than libraries, is the Internet. It allows ready access to a vast range of sources including company

Data Analysis and Presentation Skills by Jackie Willis.
© 2004 John Wiley & Sons, Ltd   ISBN 0470852739 (cased)   ISBN 0470852747 (paperback)

sites, library catalogues, newspaper, magazine and journal sites, scientific databases and academic institutions. Learning how to use this resource properly is equally as important as learning any laboratory technique or other technical skill. This section will demonstrate how to use the Internet as a research tool and as a guide to electronic communication.

The Internet works through a highly complex system of networks to which a single computer may have access by a set-up as simple as a modem linked to a phone line. The Internet, or World Wide Web (www), is searched by means of a browser. This is software that will allow various sites on the network to be visited by supplying a specific address for a website. This address, known as the URL (Uniform Resource Locator) is used to access each page of information on the web. The most commonly used browsers for Internet access are Netscape Navigator and Internet Explorer, each functioning in exactly the same way.

Initially the Internet was developed by academic communities to provide information to students to support their learning and to promote research. The Internet has now become far more commercialized as many companies and businesses have their own websites to advertise their products and services, as do non-profit organizations, schools, colleges and universities and even individuals, worldwide. Information is available as straightforward text, but web pages can also provide graphics, sound, video clips and interactive components. Web pages will also usually contain links (hyperlinks) to other sites around the world, hence the name World Wide Web. In using the Internet as a source of information it must be remembered that it is not catalogued as we would expect of a standard library and is completely unregulated for quality. Web resources should therefore be accessed carefully and assessed for their validity and usefulness.

Going on-line

The Internet is connected through a computer and a networked system or phone line by means of an Internet Service Provider (ISP). The ISP usually requires a subscription for this service and many offer a free or reduced fee for a limited trial period. You will need to use an ISP if you are linking to the Internet privately and most service providers supply the browser software for you to install and then register with them. When choosing an ISP you should consider the quality of the service that they offer (i.e. reliable and quick access when logging on, uninterrupted service) and not just think about the subscription cost alone.

To access the Internet you should double click the browser icon from the desktop of your computer once you have entered Windows. Having entered your username and password if required, you should usually arrive at the homepage of your ISP or institution, if you are accessing the Internet through your university or college. At the top of the page you will see the URL for the homepage on the Location bar. To move to a different website, type the URL into the location bar and press the Return key.

Exercise 2.1

Enter the browser used by your computer system. From the location bar, type in www.wiley.com; this is the URL for the publishers of this book, Wiley. As seen in Figure 2.1. this takes us through to the company homepage for which the URL is http://www.wiley.com/WileyCDA/. If we were then to click on one of the menu items (e.g. Chemistry) then the URL will change as a different page within the site is loaded. If we wanted to direct someone's attention to a particular page then we would refer to the URL for that specific page and they would be able to go to it straight away.

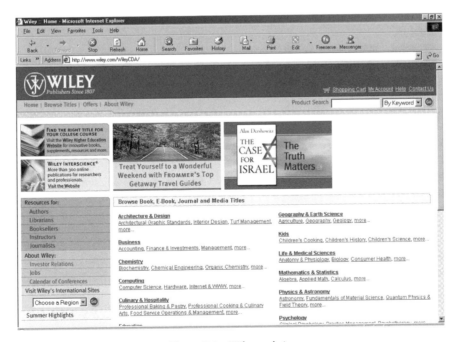

Figure 2.1   Wiley website

The information provided by the publisher is kept on a computer, known as a server, attached to the worldwide web. The homepage was successfully located because the company have a registered domain name called wiley.com. Domains are registered by organizations such as Nominet, so by typing in the URL for the domain the website on a server may be located from any world-wide location. Domain names have set conventions, which can be useful for finding information that we suspect may be located at a particular site. The .com part of the domain tells us that the web page is for a commercial organization. Some domain names indicate the type of organization and its location, for example

http://www.coventry.ac.uk

The .ac part of the domain tells us that the web page is for an academic institution. The .uk clearly shows that it is located in the United Kingdom, so looking at the URL if we had never visited the site would tell us that the domain belonged to Coventry University.

Other conventions are:

.edu   American/Australian educational institution

.gov   government (non-military) organization

.net   networking organization

.org   non-profit-making organization

Other country codes are:

au	Australia	be	Belgium	ca	Canada	ch	Switzerland
de	Germany	es	Spain	fr	France	gr	Greece
in	India	it	Italy	jp	Japan	nl	Netherlands
nz	New Zealand	se	Sweden	us	USA	za	South Africa

Knowing these conventions is useful when we are trying to look for information that is likely to be located on the web pages within a particular academic institution or company as we are able to guess the URL for the homepage. Sometimes abbreviations or acronyms are used, e.g. the British Broadcasting Corporation's home page can be found www.bbc.co.uk and America Online at www.aol.com.

Sometimes making an intuitive guess at a URL can save the time that would be spent searching through the Internet looking for a particular site.

Search engines

If you don't know the URL for a specific domain, the next step is to use a search engine. These are sites on the Internet that hold databases of millions of websites across the World Wide Web. By inputting keywords, the engines search through their records and produce a list of websites matched to the keywords that you have entered. There are many search engines available, and sometimes it is necessary to use more than one to locate the information that you are looking for. Search engines are fed by spiders or crawlers that search and index the web on a regular basis, picking up keywords from new sites that are then stored in its database. You can browse the Web using the categories provided in the search engine menus; but these are usually too generalized and are aimed at the general consumer. A more successful search will be made if you interrogate the engine's database directly with specific keywords. Where keywords are used as a phrase it will need to be enclosed in quotation marks, otherwise you will be given lists of websites that contain each word on its own. It is possible with some search engines to trace the origin of sentences that are quoted in a document at a particular website – a useful tool for lecturers marking essays (so don't be tempted to plagiarize!).

Using tools known as Boolean operators, keywords can be linked in specific ways:

AND instructs the search engine to find sites which contain all of the keywords that have been entered.

OR instructs the search engine to find sites that contain any of the keywords entered.

NOT instructs the search engine to find sites that do contain the keyword *before* the NOT but do not contain the term *after* the not (e.g. influenza NOT vaccination).

Some search engines use plus and minus signs instead of the NOT operator (+influenza – vaccination). Where these are used it is important to not leave any spaces between the operator and the keywords.

You can also use a tool known as a wildcard *. This instructs the search engine that any text could be substituted in place of the asterisk, e.g. hyper* could return documents that contain hyperhistory, hypercube, hypertension. To find out which operators are supported you should check the individual search engine pages to ensure that they can be used. The help pages or FAQs will also usually give guidance on these features.

Once you are ready to start your search you should enter the keywords or phrase in the dialogue box on the chosen search engine. You will then be given a list of 'hits' where each hit is a hyperlink through to the website that contains the keyword of interest and a short description of the site (usually the sentence on the site in which your keyword appears). Using a search engine cannot guarantee success. If your keyword is in common usage you may get thousands of hits, and therefore have to refine your search to narrow down the field. Alternatively, the search may yield none or very few results. The key to success is to be adaptable and be prepared to think of alternatives for the keywords that you are using. Search engines all have their own characteristics and allow different kinds of queries to be made. By trying different ones you will soon find one or two that are more successful at providing information you are looking for. Ideally you should settle with one that has a large database, is updated on a frequent basis and from which you are able to quickly access information. Some search engines give the time it takes for the search to give results. One of the most popular search engines for accessing scientific information is Google. Try the searches given in Exercise 2.2 using Google as a demonstration on how to use some of the Boolean operators to make successful searches. The number of hits at the time of writing is given, but this will be different when you try it as the database on the search engine is updated continuously.

Exercise 2.2

From your browser window go to Google by typing in www.google.com on the location bar. We are going to conduct a search to find out what the treatment is for anthrax in animals. If we simply enter anthrax as a keyword then a huge number of hits will be obtained. As can be seen in Figure 2.2 there were a total of 1 710 000 hits for anthrax alone, far too many to check through. Looking through the list of hits shows there is a great diversity in the returns that have been made. The first three concern bioterrorism and the fourth item on the list refers to a rock band called Anthrax and so has nothing to do with the disease. To make the search more focused, additional keywords need to be included. This time try:

ANTHRAX TREATMENT ANIMALS

(Note: the operator AND has not been used between the keywords as Google automatically searches for all keywords in the list.)

This makes a return of 49 400 hits, so although the search has been narrowed down there is still room for improvement as the top hits are still directed at the treatment of human subjects following a bioterrorist attack. This time repeat the search including the minus (−) operator to exclude any hits that give results for bioterrorism.

ANTHRAX TREATMENT ANIMALS−BIOTERRORISM

This time the search gives a list of 12 000 hits. The one at the top of the list provides information on the cause and treatment of anthrax in animals and so provides the information that we were looking for. This shows that by using keywords and filtering out the problems in the information that is being sent back to us, we are able to refine the search so that it becomes relevant to our subject of interest.

Google has other useful features. By selecting Images on the homepage, we can restrict a search to finding pictures and photographs. The Language Tools option allows searches to be directed to a specific country or for results to be shown in a chosen language.

Other popular search engines include:

Lycos (http://www.lycos.com)
One of the earliest search engines to appear on the web, but is consistently updated and still one of the most popular engines.

Ask Jeeves (http://www.ask.com)
Questions may be posed to this search engine which portrays the character of the fictitious character Jeeves, the knowledgeable butler created by P.G. Wodehouse.

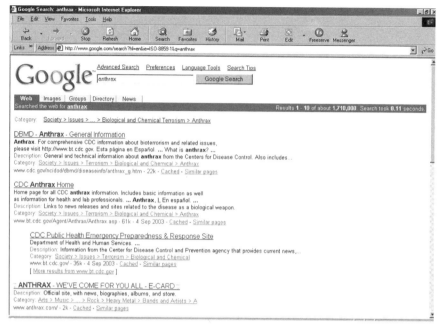

Figure 2.2   Using Google for searches

Search gateways

Search gateways provide databases that have been vetted and entered manually as opposed to automatic web crawlers that access keywords on web pages used by the search engines. The most established gateway is Yahoo (www.yahoo.com). Unlike search engines they are often reviewed for the quality of the information they hold, so the databases are more likely to contain keywords that are highly relevant and supply hits of good quality. Clearly reviewing all of the web manually is a very time-consuming and limited process, so it is possible that some sites could be omitted.

Keeping a record of information from searches

When information that we find on a website is useful we usually want to keep a copy. We have a number of options available by which this information can be stored. The most obvious one is by printing out the information. Some web pages have very elaborate borders and may contain advertising banners. If we are only interested in printing the text on the page, then in the Print options (from File: Print) choose 'Print only the selected frame' from the list. You can of course choose to print the whole of the page or individual frames of the page, but either of these will be time-consuming and probably unnecessary.

If we want to collate some information on a given topic then it may be more practicable to paste some of the information into a Word document that we can add to as we go along. One issue here though is that of copyright. Any information supplied on the web is covered by copyright laws in the same way as written text in books, journals or newspapers. Text may be highlighted and then copied and pasted into a Word document; if you do this then make certain that you also make a note of the source of the item that you have duplicated. It is important to acknowledge the source of all information to avoid any possibility of plagiarism.

If you want to keep a record of all of the information on the web pages then you could save the information on disk. By clicking on File: Save As and entering a filename you will be able to keep a record of the page as an html file that can then be opened in the browser window and will therefore maintain the full features of the web page including any diagrams and photographs.

Making effective searches

Learning how to use the web effectively is the same as using any other resource. Keep the following points in mind before starting:

- Make sure you know what you are looking for before you start and don't allow yourself to become distracted by irrelevant information (which can be hard if it is interactive or appealingly presented).

- Write down keywords (particularly important if you use more than one search engine – it's easy to forget what you have already covered) and be prepared to adapt them.

- Try to identify specific organizations that are involved in the area that you want to research. Sometimes they may have a resource on their site of the information that you are searching for. Many academics use their institutional website to display slides used in conference presentations or lectures and lists of their publications.

- Keep track of your movements. If you find a particular site of interest, then use the bookmark or favourites facility to keep a record of the URL for the page so that you can easily return to it at a later date. This is accessed by clicking on the Bookmarks or Favourites button on your toolbar, adding the URL to the list of sites that you want to keep on file.

Making searches can also be time-consuming if you are having to access the Internet during the business hours of the United States, currently the main user of the Internet, which will be peak time for its use. Loading times for the main search engines and websites are noticeably slower during this time due to the heavy traffic on the web. To help speed up these processes, use the following tips.

Try clicking the Stop button if a page is downloading very slowly. This usually shows the text without the graphics. Some websites allow the user to load the text-only version of the website to disable the loading of those elements that load slowly.

If you want to follow a hyperlink from a site, click on this as soon as it is visible. It is not necessary to wait for the whole of the page to load.

Meta search engines

Meta search engines search all of the databases from search engines for you, sometimes making it easier to access information than by trying individual search engines. This approach is worth trying where you are getting only a few or no hits with your keywords. Try using Metacrawler (www.metacrawler.com) for the searches that were made earlier in Exercise 2.2. You may find that there are fewer hits, but the results are more relevant.

Organization and institutional websites

Various professional associations and societies and the media all have web pages which they are keen to promote. If your interests are focused in one of these areas or covers a topical issue it is usually worthwhile checking the relevant society website. A list is provided on the resource website for this book of some of the most useful ones in different subject areas.

Copyright and validity

Material found on the Internet is there for public access. You must not, however, download or copy information as though it were your own original material. It should be considered in the same way as items from a textbook or publication, i.e. you should quote the author and source of the information if you are referring to it elsewhere. Before accepting the content as valid you should make every effort to find out its origin, i.e. the scientific publication from which the author of the website took the information. If you are certain that the website is valid, but cannot find the original source then your reference should include a title, author and the URL for the web page, as shown in the example below:

Fletcher, G & Greenhill, A 1995, Academic Referencing of Internet-based Resources, Australian Library Journal, http://www.spaceless.com/WWWVL/refs.html

If you are unable to find anything to verify its validity, then you *should not use it*!

Searching for scientific publications

In order to be able to perform a search of scientific research papers worldwide you need to use a powerful database. Perhaps one of the best-known databases is Medline which contains details of medical research publications; this database is usually accessed from a larger database of scientific publications. The Web of Science ISI Citation Database is one such database, but a subscription is required and access is usually confined to a computer on campus from the subscribing institution. Using the Web of Science database it is possible to search more than 8400 of the worlds most prestigious journals, books and conference proceedings through three multidisciplinary databases: the Science Citation Index Expanded, the Social Sciences Citation Index and the Arts & Humanities Citation Index.

The database contains only those journals that have met specific selection criteria and have undergone peer review. It is updated on a weekly basis and contains scientific information collected from 1981 onwards. Your search may be confined to a specific time period or only that part of the database that has been updated since the last time you conducted a search. If your institution subscribes to Web of Science then you will need to obtain a username and password to access the site.

Available completely free of charge, however, there is another database, PubMed, which is accessible from any location. This database is a service of the National Library of Medicine in the United States. It also contains citations from Medline and other life science journals dating back to the mid-1960s. Whichever database you use, the search options are common to both. The use of PubMed will be explained here as it is available to everyone.

To access PubMed use the URL: http://www.ncbi.nlm.nih.gov/entrez. On the home page there is a general search box in which you can perform a quick search, but it is usually better to click on Limits. This will allow you to refine your search, perhaps to specific dates or for publications printed in English. In the general query box (at the top of the page) you are able to search using any of the following terms:

- Keywords associated with your search (with operators AND, OR, NOT in uppercase).

- One or more authors of a scientific paper. The last name is entered first then initials, but without any punctuation, e.g. willis jv. If only the author's last name is available then change the options under Limits (where the default is All Fields) to Author otherwise all available fields will be searched which could give even more returns for an entry such as 'brown' (you can try this for yourself and see the difference).

- The source of the article, i.e. the scientific journal in which the paper was published e.g. Molecular Pharmacology.

- Date of publication of the article, which would allow checking for only recent publications, or to limit viewing articles from a particular journal or author in a certain year.

Once a search has been made there will be a list of hits returned, showing the title, full list of authors and journal details. For many items there is an abstract of the article and the facility to view related articles to a particular scientific paper. For some titles there is the provision to link directly to the publisher to gain access to the full text of the publication. This information can be printed directly from the web pages. For some papers this service will be free, but for many a fee, payable on-line using a credit card, is required. (Some medical journals may be accessed completely free of charge for specific years; these may be found on www.freemedicaljournals.com).

When you have completed a search, the results may be copied to disk or sent to your e-mail account by selecting the appropriate option from the Send To menu at the top of the page. This is particularly useful when a large number of hits are displayed and you need to take time in looking through them. The results of a search are normally displayed on a page of up to 20 items through a number of pages which are accessed using the Next or Page button at the top of the web page. The article appears as a summary, but an abstract (where available) can be displayed by clicking on the list of authors.

Electronic mail

The Internet is also used for the electronic transfer of documents, known as electronic mail, or e-mail for short. E-mail allows the instantaneous sending of messages anywhere in the world and by attaching files to e-mails we can easily transfer reports, photographs and presentations in a matter of minutes. It's hardly surprising that the use of the traditional postal service has earned the name 'snail mail'. E-mail is usually available as part of the browser software. The main packages are Outlook Express and Netscape Communicator (known as e-mail clients). Details of setting up an e-mail account are not provided here as these will vary with the software used and the ISP that you are connected to, and some e-mail accounts can be set up on the Web that can be accessed from anywhere. These are offered by sites such as Hotmail (www.hotmail.com) or Yahoo (www.yahoo.com). Instead we will go through the basics of receiving and sending messages that are in common with whatever version of e-mail you are using.

Whichever way you are connected to the Internet, you will have a mailbox that is located on the server that is hosting your e-mail account. To access your account you will need to double click on the e-mail icon if you are using this from your desktop. Before opening your mailbox you are prompted to enter your password (and possibly your username). Once this has been accepted you are then faced with a split screen that shows:

1. a mailbox which has an Inbox into which your incoming messages are received and the Outbox from which any messages waiting to be sent will be kept. The Inbox may have a number of different folders for Received Mail and Sent Mail. Clicking on either folder will show its contents.

2. messages that have come into your Inbox. These are listed in date order and show the name of the sender, subject of their e-mail and the date and time at which the message was sent. By clicking on a message it will be opened for you to read.

In order to see how these functions work you must first send a message of your own.

Exercise 2.3

Composing a message.

Click on the New Msg/New Mail icon (near the top left of the screen).

A dialogue box appears into which you need to type the e-mail address of the person you want to send an e-mail to. For example, if you wanted to send a message to someone with an e-mail account at hotmail, then the address would be their username@hotmail.com. When you sign up for an e-mail account then you are asked to choose a unique username for yourself to identify your account. To make sure that you are able to use e-mail properly, send the first message to yourself.

In the address box type: username@hotmail.com where the address is your own e-mail address that you signed up for.

Next you need to complete the e-mail subject box. This needs to be a brief title to indicate the content of your e-mail (equivalent to writing 'Re: . . . . . .' in a letter).

Type in: 'My first e-mail message'.

When you are ready to send your message, click on the Send button and the e-mail will be sent.

E-mail etiquette

Although e-mail has become a more informal way of communicating, you should always adopt the same basic rules as you would with a written letter. Do not use WORDS IN CAPITAL LETTERS (as shown). In e-mail etiquette this is the equivalent of shouting at someone, and so you will cause offence if you do this.

E-mails have also become the means by which computer viruses are transmitted and you must be alert to this possibility. A lot of junk mail is sent via e-mail (known as spamming). If you see a message from someone that you do not know, then delete the e-mail without opening it. If you receive a message in which capital letters have been used throughout the subject heading, then again do not open it, but delete it straight away as this is one of the indications that an e-mail may contain a virus.

If you receive spam that requests you to return a message asking them to remove you from their list, then ignore this as it is usually a ploy to add your details to their address database. Any e-mails that you suspect are fraudulent, e.g. offering money in exchange for transferring funds from a foreign bank account into your own account, should also be ignored and deleted, as these are from criminals who want your bank account details in order to falsely take money from you.

By now the e-mail you sent to yourself should have appeared in your Inbox. Double click on the subject to open the e-mail. If you wanted to make a reply to the message, then you could click the Reply button and type in a message above the one just received. You also have the option of sending the message to someone else. If you click on the Forward button you can then enter the e-mail address of the person that you want the message sent to and then click Send.

You are able to view a copy of the message that you sent by moving into your SentMail folder (on entry into the e-mail client you are automatically taken to your Inbox to see new messages).

Click on Local Mail, then SentMail.

SentMail shows all messages that have been sent by you and the e-mail address of the recipient. By double clicking on an item you may open the message and read it. As your personal allocation of space on the server that holds your e-mail is likely to be limited, you should routinely check through messages in your SentMail and Inbox and delete any old messages that you no longer want. A message can be deleted by clicking on the item to highlight it and then pressing the Delete button on the keyboard. The message will then be sent to the Trash, which will be emptied when you close the application.

If you want to keep the e-mail address of someone who has sent you a message then click on the Tools option and then select Add Sender to Address Book. Their details will then be automatically entered into your virtual address book. When you next compose a message by clicking on New Msg you will be able to find their address by typing in their name or by browsing through the list. Once you have found the address that you want, you click on the address and then To: and it will be pasted into your message. You may paste as many e-mail addresses as you want into the message, so you can e-mail several people at the same time.

Opening and sending attachments

E-mail is a convenient way of transferring information rapidly. Files may be sent that contain data, text, photographs, clip art or even music. Some lecturers send copies of lecture notes or handouts to students via e-mail, so it is important to learn how to send and open these documents.

Exercise 2.4

Find a document that you have created in Word (or create one now) and save it to your A: drive on a floppy disk. Compose another e-mail addressed to yourself with the subject heading 'Attachment'. In the main body of the message type a short message and then click on the icon that will look like a paperclip, which is for sending an attachment. When you select this option a browser window will appear in which you will be able to select the file you created from the A: drive and attach this to the message. Once you have done this, send the message and then wait for it to appear in your Inbox. When the message arrives it will be shown as having an attachment by being marked with a blue paperclip.

Open the message by double clicking on it. Find the attachment by looking at the bottom of the message. It should appear in a separate box with the filename written on it (or may be shown as a file at the top of the message). Double click on the filename to open it. A dialogue box may appear indicating what application should be opened in order to view the attachment (in this case Word) and give you options to either open the document now or save it to disk.

Choose the option to open the document.

Word will now automatically load and the document will open.

Discussion groups

Once you have set up an e-mail account you will be able to communicate with other people. The Internet has developed 'communities' where people with common interests may have discussions and share ideas. These take the form of:

- Mailing lists. These are semi-private discussions which take place between a select group of people, usually specialists in a particular field if they are an academic mailing list.

- Newsgroups for public discussions with free access by anyone. These cover a range of different subjects: hobbies, sport, music, art, cooking, etc.

- Chat groups (held in virtual chat rooms) for real-time discussions that can be public or private. There have been a number of on-line conferences held by the scientific community in recent years where these real-time discussions have taken place.

Mailing lists

Mailing lists make use of e-mail to allow a conversation with a group of people. Thousands of mailing lists exist on the Internet. You need to find which mailing list you are interested in joining and then subscribe to it. Subscribing means that you register yourself as a member, but there is no charge for joining the list. Whilst you are a member you will be automatically sent messages from the list to your e-mail address and you will also be able to send messages. If you decide at a later time that you no longer wish to be a member then you unsubscribe and your details will be removed.

As scientific topics tend to be of interest to specialist groups, you are more likely to find relevant information and discussions taking place in a mailing list than a newsgroup. One of the most widely used resources for mailing lists is JISCmail.

Go to http://www.jiscmail.ac.uk. From the Find lists menu select Biological Sciences from the drop down list under category pages and then press Go. You are then provided with a list of groups, each with a short description about the purpose of the group. If you click on the title of the group you are then presented with a list of all the weeks in which messages have been posted. By clicking on the date at the top of the list (most recent) then you will be able to read through the topic currently under discussion by the group. Another popular resource is LISTSERV that may be found at http://www.lsoft.com/lists/list_q.html.

Once information has been researched from literature reviews and Internet searches, we are able to start formulating ideas. This then takes us through to the planning stage where we start to consider the design of our investigation.

2.2 Experimental design

Having thoroughly researched the background of a scientific topic we begin to formulate our own ideas. From these we produce hypotheses in order to explain any unanswered questions. In the statistical analysis section of this book we will be looking at how to test our hypotheses, but before that we have to plan an experiment or investigation to answer the questions raised by our hypothesis. An important part of the planning process involves the correct design of an experiment. We are going to briefly review some of the issues in experimental design, but a wide range of examples of experimental designs appropriate for different types of investigations are introduced in later sections of the book when statistical analysis is also considered.

Planning an investigation

There are a number of important steps in planning a successful investigation, whether this is a laboratory experiment, clinical trial or fieldwork exercise. In each case there are key points that it is important to address. Whatever the type of investigation the experimenter should:

- establish the objectives of the investigation. What is the question that you want answered, are your hypotheses sound and are you certain that you can achieve the results that you are seeking?

- determine the size and characteristics of the sample that you are going to take. Is this realistic in the planned timeframe and how will you select test subjects?

- choosing the methodology. Is the experiment unbiased and will there be appropriate precision in the methods used?

- select an appropriate design and plan any statistical analysis. Will the experimental design allow statistical analysis and if so, what test do you plan to use? Are sufficient subjects or replicates included to make this viable? Is a pilot study necessary to pre-test an aspect of the investigation?

Establishing aims and objectives

Before starting to plan an investigation it is worthwhile considering what you are trying to achieve in doing the experiment and making sure that the objectives which you set are realistic. You may have formulated hypotheses by conducting some background reading and then come up with your own ideas. It is always useful to discuss your plans with someone such as a tutor to make sure that the ideas on which they are based are completely sound and not flawed in any way. If the objectives of the investigation are unclear then the results obtained will not help to resolve the research question posed. It is important in the early stages of planning to make sure that your objectives are clearly established.

Populations and sampling

The collection of data gathered during an investigation is called a sample; the sample is just a small part of a (much larger) population. The population can be any living organism, e.g. it could be plants if we were studying the heights of a particular species of tree within a given area; or it might be the size of isolated cells measured under the microscope. Sometimes we may look at more than one population in an experiment, for example, we might compare the sizes of cells from different organs of the body.

In the biological setting, populations are in a state of continuous change, for example organisms develop, grow and reproduce and some may die or become diseased. It is clearly impracticable to collect information from every member of a population so we have to limit ourselves to what may be practically and realistically achieved and take what we expect to be a representative sample of the population. The question then arises, how large a sample needs to be obtained to be truly representative? Logically large samples will be more representative than small samples, but constraints of time and money often limit the size of a sample that can be made.

The purpose of the investigation must be carefully considered when deciding how samples are going to be taken. It is usually helpful to look at previous investigations similar to your own to see what size sample was used and whether the investigators demonstrated that sufficient numbers were taken to represent the population.

Sampling may be either random or proportional. In random sampling every member of a population has an equal chance of being selected for the sample, so there are no special limits applied to exclude certain members of the population. Alternatively we may wish to use proportional sampling where the sample needs to be representative of an aspect of the wider population and so we have to be selective about including subjects in the sample. Breast cancer is a disease that occurs mainly in women; it does occur in men, but very rarely. If we wanted to examine the genetic predisposition of individuals for the disease then it would be clearly inappropriate to use a sample that contained a large proportion of male subjects.

In deciding how large a sample to obtain we also need to consider the magnitude of the difference that we are looking for in our experiment. If we are expecting to see a large difference when we compare two samples then a small number should be sufficient to demonstrate an effect. If we expect that there is likely to be a very narrow margin in the differences between samples then a much larger sample needs to be taken. This will also be the case where there is likely to be a high variability in the factor that we are measuring.

One rough measure of sample size can be determined from using a running mean. If we were conducting an experiment in which we were measuring a particular variable, for example, the height of conifer trees that have been grown for one year, we might start by taking a sample of 10 conifers from which we would calculate the mean height. This gives a value of 97.9 cm. Each time we made a further measurement we could recalculate the mean; this is known as the moving or running average. The moving average for a further ten samples is shown in Table 2.1. If we plot this data then we can see that the moving average has settled on a value of about 98.2 cm, so we can be reassured

Table 2.1   Heights of conifer plants grown for 12 months in identical conditions

Conifer height (cm)	Number in sample	Mean (cm)
113.4	11	99.3
99.7	12	99.3
98.6	13	99.2
91.1	14	98.7
87.8	15	97.9
104.2	16	98.3
95.5	17	98.2
98.1	18	98.2
92.4	19	98.0
100.4	20	98.2

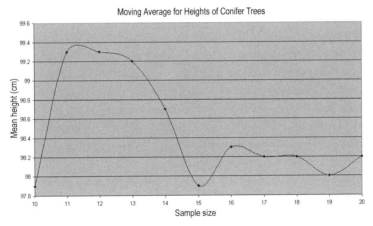

Figure 2.3   Graph showing moving averages for the heights of conifer trees

that there is no need to take any further measurements as the sample size is sufficient. The moving average can be seen in Figure 2.3.

Although this approach may be useful for determining sample size in some situations, when we design an experiment or investigation we usually need to estimate the sample size before the start of the study. This is due to practical reasons such as the length of time that will be required to perform the investigation and because costs of research need to be justified before a study may commence. There are a number of calculations that can be performed to calculate the required sample size, the details of which will not be entered into here. The support website for the book gives some examples of how sample size may be calculated and provides links to on-line resources for calculating sample size.

It is important to establish the correct sample size for an investigation. If the sample size is too large it may be unrealistic to conduct the investigation or may prove unnecessarily wasteful and costly, particularly where a smaller sample size could have been used with some modifications to the experimental design. If the sample size is too small, any differences that should have been demonstrated may fail to be identified. Statistically this is known as a Type II error: the sample taken is unable to demonstrate any effect that could be identified by a statistical test. This is distinct from a Type I error in which a statistical difference is shown but cannot be justified (for a further explanation of this phenomenon see section 5.3).

Choosing methodology

If the investigation is going to be laboratory-based then there is normally an analytical method that needs to be set up and validated for the experimental conditions to be investigated. Factors such as reproducibility of results, the precision of the method used and limits of detection need to be determined. Once these are established, the way in which the experiment is planned will help to minimize any variability in the data and help to ensure that a fair test is conducted. In many biological experiments a control group is a common feature incorporated into experimental design and is included to prevent there being any bias in experimental results.

A control is a group of subjects or series in an experiment to which no active treatment is applied. For example, if we were conducting a laboratory experiment in which we wanted to investigate the effect of ultrasound on the disruption of bacteria, then we would include a control in which the ultrasonic probe would be placed in the bacterial suspension for the predetermined time period, but no ultrasound would be applied. This would provide exactly the same test conditions as for the samples exposed to ultrasound, eliminating any effects associated with the mechanics of manipulating the bacterial suspension in this way. Where we are dealing with human subjects we have to take different measures to produce a control as humans are cognitive and can sometimes influence results, either deliberately or subconsciously. In trials involving human subjects, the design has to be controlled carefully to prevent any bias being incorporated into the experiment by either the test subjects or the investigators themselves.

If we wanted to test a new drug for its analgesic (pain-relieving) properties then we might ask a test subject to rate how they felt during the course of a day, assuming that they suffer from a condition in which they experience chronic pain that requires continuous medication of this type. Clearly we have to give them the test drug and most likely this will be in tablet form as this is the most common form of administration. If we ask the test subjects to compare their degree of pain with a period in which they are provided with no medication whatsoever then we would immediately create bias in the experiment, as patients would automatically assume that without any painkillers they are likely to experience far more pain. How then are we able to conduct a fair test in this situation? The answer is to use a placebo (the literal translation being 'I shall please'). A placebo would be a dummy tablet – one that is made up in exactly the same way as the test tablet but without any active ingredient; the test substance is likely to be substituted by a filler such as talc or sucrose. The tablet is then given to the subject in the same way as the active tablet. If we were to record the pain experienced by the patient then this should provide us with a fair comparison – but providing we have not incorporated any bias from other sources. Firstly we must make sure that the participant in the study is unaware of when they are receiving the new drug or the placebo. This may be done by randomizing the study so that for some subjects the placebo will be taken before the test drug and vice versa. The study conducted in this way would be single blind as we would not reveal to the patient in which order they were taking the two treatments. How though could we prevent any bias incorporated into the study by the investigator? If they are aware of the order of treatment, they may feel guilty at providing a patient with a placebo that they know will have no effect and perhaps enthusiastic when giving the new treatment as they might expect some interesting results. The way round this is to make the study 'double blind'. A randomization code is set up by an independent third party who then assigns the tablets to be used in the study so that neither the investigator or the test subject is aware of which treatment is being taken. The code would only be broken in the event of a serious adverse reaction by the patient, in which case the clinician would need to know what was being given in order to apply counteractive treatment if necessary. In using placebos in clinical trials, various ethical issues also have to be considered. Firstly, under what circumstances can the use of a placebo be justified. Where the well-being or life of a patient is endangered then a placebo would be inappropriate. Under these circumstances a standard treatment has to be compared with a newer test substance. Ethical committees approve all protocols for clinical trials and it is part of their duty to ensure that the study will not be detrimental to the patient in any way.

Sometimes having planned an experiment it is useful to do a test-run or pilot study to ensure that all of the conditions selected for the experiment are appropriate. In an experiment intended to compare drug dissolution of tablets with different properties, we would waste valuable resources by sampling the dissolution medium every 30 minutes for a period of 10 hours, with replicates of 10 tablets, if we then find out that the dissolution process is complete in 20 minutes. Our background research should, of course, have indicated this, and it is an extreme example; but we should always keep in mind that science has previously shown a trend for demonstrating the unexpected, and that when the unexpected does occur then there is usually something very interesting to follow up – but we have to follow it up in the correct way. Our experimental design is crucial to this process, and sometimes we need to be adaptable and rethink our ideas to include factors we may not have previously accounted for.

WEB SUPPORT: SECTION 2

2.1 Sources of information

Websites are always continually evolving, organizations change names and new material is to be found. URLs are not listed in the book for this reason. Instead you will find many useful links to journals, professional bodies, electronic databases and institution websites on the Support Web. This will be updated so that as new resources become available you will be directed towards them.

2.2 Experimental design

Here you will find some links to useful sites with which to estimate sample size for an investigation. There are also examples of different experimental designs (for laboratory-based studies, fieldwork and clinical trials).

3

Presenting Scientific Data

Once we have completed an investigation we are confronted with a wealth of information that needs to be summarized, analysed and evaluated. One of the first tasks is to collate results and present them in the form of graphs or charts, having calculated some basic statistics such as the mean and standard deviation. In this section you will learn how the software application Excel can be used for summarizing and presenting data.

3.1 An introduction to Microsoft Excel

Excel is a software program that uses spreadsheets organized into workbooks. A spreadsheet is an electronic worksheet composed of individual cells arranged as a grid of rows and columns. Each cell can contain data or a formula used for calculations from information in specified cells. Excel is used for a variety of purposes ranging from simple calculations to statistical analyses and producing charts and graphs, and even as a database. In this and following sections we shall be exploring the use of Excel for these functions, but we will make a start by finding out how a spreadsheet is organized and used.

Data Analysis and Presentation Skills by Jackie Willis.
© 2004 John Wiley & Sons, Ltd   ISBN 0470852739 (cased)   ISBN 0470852747 (paperback)

Figure 3.1   The Excel spreadsheet

Setting up a worksheet

Once Excel is opened a blank worksheet appears on your screen as shown in Figure 3.1. You will see that the columns of the worksheet are labelled alphabetically from left to right and that rows are ordered numerically. Each cell will therefore have its own unique reference such as A4 or D20 which can be used to group data or identify the cell in which particular values may be found. At the bottom of the screen is a tab showing which worksheet is currently open to work on (just like a subject divider in a folder). In a new workbook the default sheet is Sheet 1. By clicking on the tabs you can move from one worksheet to another.

Sheets can be moved, copied or rearranged within a workbook and between workbooks and each sheet can be given its own customized label. In the following section we will learn how to use these functions in Excel.

Entering data

Open up a new workbook by entering Excel. If a spreadsheet is not automatically opened press File|New and a new sheet should appear. Although there are many cells on the spreadsheet there will be only one 'active' cell in which you can enter information. This is identified by the heavy border surrounding the cell. Any information that is entered into the cell by typing information from the keyboard will be displayed in the cell and, while the cell is active, on the formula bar located above the spreadsheet. This is an important feature as some of the equations that may be entered into a cell can be quite long (and subject to being mistyped) so it is essential to be able to read these in full on the formula bar to check that they are correct. After entering the information, pressing the Return button will enter the contents into the cell; pressing the Delete key will remove the information. Data can be inserted into worksheets in either rows or columns, but many of the functions applied in Excel usually work best if the data is entered into columns. We will use Excel to input the information in Exercise 3.1 and then apply some of the features of Excel to produce basic summary statistics and a chart.

Exercise 3.1

Copy onto the worksheet the information shown in Figure 3.2 (keeping cell references the same). If any of the text disappears into the cell when you have entered the data, just ignore this as you will be editing later.

You are now going to enter some data for May, but to do this you are going to use a feature known as the Autofill function. Move to the cell containing the label for April (this should be D4) by clicking on it and so making the cell active. If you move the mouse pointer around the border of the cell you will see a white arrow appear, but when the bottom right corner of the cell is reached (on the small black square located there) this changes to a cross. The black cross indicates the Autofill handle is active. Keep the mouse button pressed down and drag the Autofill handle through to cell E4. As the cell is dragged across the row a label for May appears in E4 (you could continue dragging through other cells and calendar months would automatically be entered into the cells – try this out as they can easily be deleted afterwards). This Autofill feature also works for days of the week and numbers in a series, as Excel is able to recognize trends in data entered on the sheet. For instance, where two adjacent cells are highlighted that contain

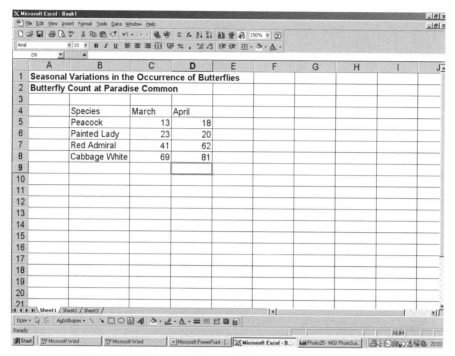

Figure 3.2   Entering the butterfly data into the spreadsheet

two numbers in a series such as 3 and 6, the Autofill will continue the series by automatically entering 9, 12, 18 and so on.

Complete the table by entering the remaining data on the spreadsheet. To make this easier, select the cells where the data will be entered, i.e. select cells E5 through to E8 (usually written as E5:E8) by highlighting them using the mouse. Then type in the following values:

Peacock   20
Painted Lady   16
Red Admiral   99
Cabbage White   91

Functions in Excel

There are two useful functions available from the Excel toolbar; these are the AutoSum and Paste Function. In some versions of Excel the Paste Function is

Figure 3.3   AutoSum and Paste Function (Function Wizard) buttons

called the Function Wizard. These two tools allow mathematical and statistical functions to be applied to data on the spreadsheet. By using these features you are able to perform complex calculations with the click of a button. Figure 3.3 shows these two function buttons that are located on the standard toolbar.

Using the AutoSum Function

Using Excel you are going to calculate the total number of butterflies observed during each month. Firstly select cell B9 and type in TOTAL. Now make C9 the active cell. Click on the Autosum (Σ) button twice. The total of the numbers in cells C5:C8 will appear in cell C9. (The package searches around the cells by C9 and recognizes that the data in cells C5 through to C8 are to be added. On the first click of the AutoSum button a formula appears on the formula bar indicating $=SUM(C5:C8)$, in other words the data in these cells are to be added. Clicking on the AutoSum button again confirms the formula and so the total appears in column C9. Pressing the Return key would also confirm the formula. If the range of cells to be totalled needs to be changed then the cell references may be altered on the formula bar or by selecting only those cells that you want to be added together.)

Formulae are identified by the $=$ sign. The formula is usually only seen while you are editing a cell that contains the formula. Within the cell the numerical value resulting from the formula will appear. Formulae can be entered manually so you could type $=SUM(C5:C8)$ or even $=C5+C6+C7+C8$, but clearly the latter approach is more clumsy, particularly where you might want the sum of a long list of numbers. Excel offers a range of functions (where AutoSum is just one example) to enable you to use a whole range of mathematical and statistical calculations very easily.

Copying formulae

The formula in a cell may be easily copied across to adjacent cells by using the Autofill handle that you used earlier to insert the months of the year. If C9 is made the active cell and the Autofill handle selected and dragged through to E9, the formula from the Sum function is modified and copied into each of those cells, so the totals for April and May will be placed on the worksheet.

This is a useful function that will be used again when making more complex calculations in Excel worksheets.

Adding text to the worksheet

Labels for columns can be added to the worksheet to form tables of data, but it is always useful to add titles and comments to the worksheet for easier reference. Text can easily be typed into individual cells, but the width of the cell will determine how much of the text is seen. The width of the column may be altered, but this may then interfere with the formatting of the table on the worksheet. The easiest way of adding text is by using a textbox. Click on the Draw icon on the standard toolbar and the function buttons that can be used in this utility will be displayed at the bottom of the screen. Select the textbox button and then by clicking and dragging the mouse you will be able to draw a textbox on the worksheet. To add the text simply type in your words and these will appear inside the box. The textbox may be positioned anywhere on the worksheet by selecting and dragging the box.

Widening/narrowing rows and columns

The width (or length) of columns can be altered by dragging its border with the mouse. The width of column B may need to be increased to accommodate the label Cabbage White. Using the mouse, move the cursor to the line separating column B from column C. The cursor sign changes to a double arrow. Click and drag the border of column B until it is wide enough to show the full names of the species of butterflies. The same procedure can be applied to decrease the width of the column or row. The width of a column or height of a row can be automatically adjusted by double clicking on the border between adjacent cells.

Managing lists of data

Data in a spreadsheet can be managed as a list. Excel has several facilities for managing data in lists; these involve sorting and ranking the data. Firstly rearrange the butterfly data by sorting through the list to see which butterflies were most prevalent in May. This will involve ranking the list numerically in descending order. Select all of the data about the numbers of butterflies from

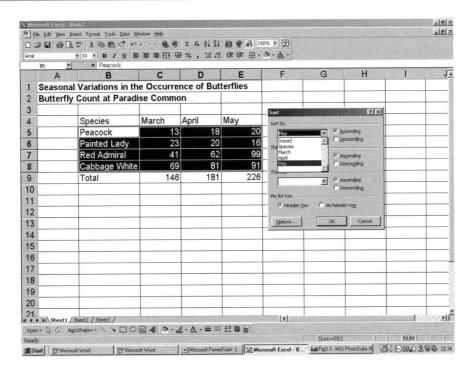

Figure 3.4   Sorting data on the Excel spreadsheet

March to May including the labels, but not the totals (i.e. cells B4 across to E4 and down to E8).

From the Data menu choose Sort. A sorting list is suggested after Excel has reviewed the data on the worksheet as shown in Figure 3.4. In the **Sort by** box type 'May' and then press the **Descending** option button; click on OK. The list will now be sorted showing the largest numerical value at the top of the list (in cell E5). By looking at the list it is easy to see which type of butterfly is prevalent in May. Repeat the process for April and sort the list in ascending order for March. By arranging the sorting order differently we can examine the relative prevalence of butterflies in each month.

Using the Paste Function (Function Wizard)

The Paste Function allows various functions to be calculated, statistical and mathematical, by automatically entering a formula into a cell.

Operators in Excel

In order to use Excel for formulae, we need the standard operators in mathematical equations. Some may be used by inserting characters from the keyboard, others may be accessed from the Paste Function menu. A list of operators is provided below:

Mathematical symbol	Excel equivalent	Paste Function
$+$	$+$	SUM
$-$	-	
$\times$	*	PRODUCT
$\div$	/	
n^2	^2	
$\sqrt{n}$		SQRT

If the contents of cells A3 and A27 are to be multiplied together then we should enter an equation onto the spreadsheet thus:

$$= A3 * A27$$

But if we wanted to multiply all of the contents of the cells between A3 and A27 then we would use the PRODUCT function:

$$= PRODUCT(A3:A27)$$

Brackets
Priority is given to various operators so always use brackets in Excel as in other mathematical equations. For example, $= 27+4/2$ will give an answer of 29, whereas $= (27+4)/2$ will give an answer of 15.5.
N.B. *Small numbers*: when the value returned by a formula or function is very small Excel will express this in exponential form, i.e. the value 0.000 005 1 (5.1×10^{-6}) would be shown as 5.1E-6.

Using the Paste Function we can calculate the mean and standard deviation from the butterfly data. Firstly we will calculate the mean and standard deviation of the number of Peacock butterflies observed from April to May. In cell F4 enter the title 'Mean'; in cell G4 enter the title 'SD' (standard deviation).

Click on cell F5 (this is the cell in which we want the value of the mean to appear). The formula for calculating the mean needs to be placed inside this cell; this is done by the Paste Function. On the toolbar (next to the AutoSum function) you will see the button for the Paste Function (*fx*).

Click on the Paste Function button and select *AVERAGE* from the selections shown (note that there are different categories to help you locate each function, e.g. AVERAGE is found under Statistical. To see the entire list select All; click on OK. You are then prompted to supply the range of cells for which to apply the formula. Enter C5:E5 then click on OK. (Alternatively you may click on the worksheet and select cells C5 to E5 by holding down the left mouse button and dragging to highlight the required cells; the cell references are then automatically entered into the box.) The value of the mean will then appear in cell F5.

Now move to cell G5 and repeat these steps, selecting STDEV from the statistical menu to calculate the standard deviation. Make sure that you keep the cell references as C5:F5 (as Excel is likely to automatically enter C6:E6 which would include the mean value in the calculation).

Using the Autofill the formulae for the mean and standard deviation can be copied across into cells F6 to F8 and G6 to G8.

Setting decimal places

Usually the values for the mean and standard deviation calculated automatically by Excel are shown with too many decimal places and so this needs to be adjusted. This is very easily accomplished by using the Increase Decimal and Decrease Decimal buttons (see Figure 3.5).

Select the cells that need adjusting (F5 to G8). Click on the Decrease Decimal button. For each click the number of decimal places is decreased, rounding up or down as appropriate. Set to an appropriate level for the butterfly data. The number of decimal places can be increased by using the Increase Decimal button. These two buttons toggle the function between them, so sometimes clicking a button will not cause any change in the data or may change the numbers to a series of ##### signs. By clicking the opposite button this will cause the change in decimal places to function so the two buttons may need to be used to alter the data to the required number of decimal places.

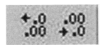

Figure 3.5 Increase Decimal and Decrease Decimal buttons

Producing graphs and charts on the worksheet

In scientific reporting we need to present data. Although tables are useful devices for this purpose a visual representation of numerical information is usually far easier to interpret than rows and columns of numbers. A graph will present a visual summary of the results and allow the observer to immediately see any trends in the data. Graphs can be presented in many different formats such as line graphs, scatterplots, bar or column charts or pie diagrams. Excel uses a Chart Wizard function that will quickly allow you to change the type of plot by the click of a button, in addition to offering many variations and editing options for the graphs that are produced. However, the final decision about the most suitable plot and the way that it is presented rests with you. We are going to plot the butterfly data to see how the Chart Wizard function works, but more detailed information on producing graphs and charts may be found in section 3.2.

Select the cells on the spreadsheet that contain both the data and the labels for the data (but not the title) – for the example provided this will be cells B4 to E8 (B4:E8).

Note: you should avoid having any blank columns in your table of data as Excel will also plot the 'empty' column in the resulting chart. Blank columns (or rows) may be deleted by selecting Edit then Delete from the drop down menu. You may then select an option to remove an entire column or row.

Click the *Chart Wizard* button on the Toolbar (see Figure 3.6). You will now see a dialogue box (Chart Wizard – Step 1 of 4; see Figure 3.7) that lists all of the chart types that you can use to plot the data.

If it is not already selected, click on Column. On the right- hand side of the box you will see a bar marked 'Press and hold to view sample'. Click on this with the mouse button and a preview of your chart will appear. You can select and view different chart types.

Choosing the 'Clustered Column' option, click Next >.

Figure 3.6   Chart Wizard button

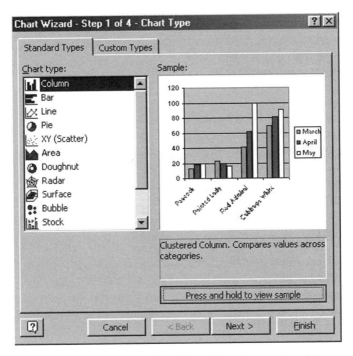

Figure 3.7 First Step in Chart Wizard showing preview of plot

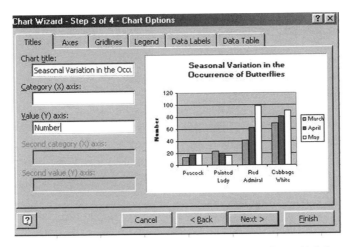

Figure 3.8 Step 3 of Chart Wizard for inputting titles and labels

In Step 2 of 4 you will now see the preview of the chart. (N.B. by clicking on the Series tab you can edit the chart, adding or removing data that you would like plotted.) Click Next > to continue.

In Step 3 of 4 you may add titles and labels to your graph (see Figure 3.8).

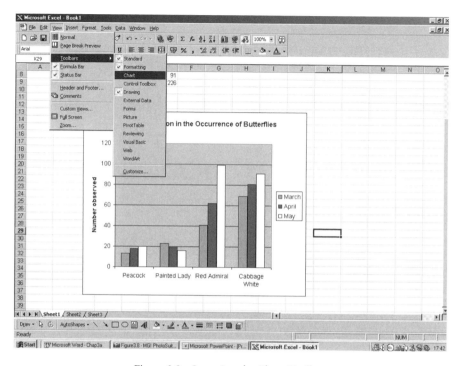

Figure 3.9   Inserting the Chart Toolbar

In the Chart title box type 'Species of butterflies at Paradise Common'. In the Value Y axis type 'number observed'. Click on Next >.

In Step 4 of 4 you have the option of either placing the chart on a new sheet or embedding it on your present data sheet. Click on 'As object in' and then click Finish.

Your chart will now appear on your datasheet together with a toolbar from which you may edit the graph. If the toolbar does not automatically appear then it can be called up by clicking on View, then Toolbars and selecting Chart from the list available (see Figure 3.9). You need to click on the graph to select it for the buttons on the toolbar to function.

Try changing the graph from being plotted 'by column' to 'by row' by clicking on the appropriate button located towards the right of the toolbar (see Figure 3.10).

Individual components of the chart can be edited using the selections from the menu on the left-hand side of the dialogue box. For instance the size of the chart can be adjusted by selecting 'Plot Area' from the drop down menu on the left-hand side of the dialogue box. You can change the dimensions by clicking on one of the black handles that appear on the border and dragging the chart to the size that is required.

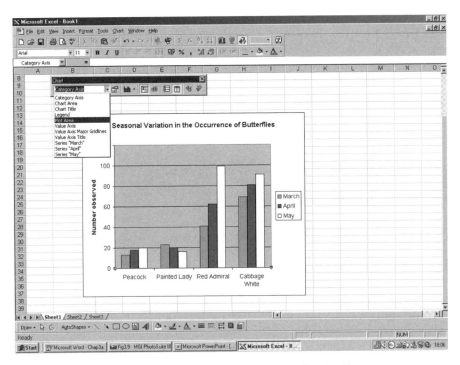

Figure 3.10 Editing the graph using the Chart Toolbar

Changing colours and patterns in the chart

Click on 'Series April' from the drop down menu on the toolbar. You should see that all of the bars relating to April have been selected on the plot (a coloured square appears on each bar). Now click on the Format Data Series button to the right of the drop down menu (see Figure 3.11). The colour palette appears from which a different colour can be selected.

Now click on the 'Fill effects' button beneath the palette and select a fill pattern of your choice. Once you have made your selection, click OK. The Format Data Series option may be used to edit other features of the graph, but this will be further explained in section 3.2.

Gridlines

Gridlines in a chart help to identify values for columns; but they are not always wanted in a chart. Click on one of the gridlines on your graph; once selected, click the right mouse button. Options appear to format the gridlines; or else, by selecting Clear, the gridlines will be completely removed from the plot. By

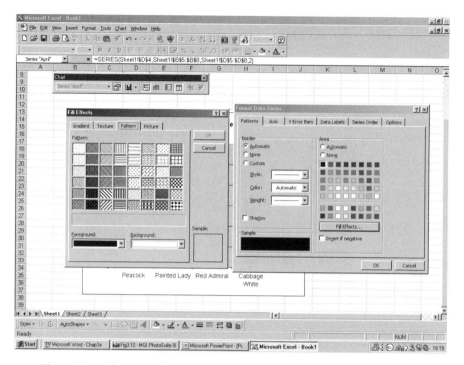

Figure 3.11   Selecting colours and patterns from the Format Data Series options

opting for Format Gridlines you are able to adjust the style, weight and colour of the gridlines.

Adding more data to a chart

More data can easily be added to a chart.

Select cells F4 to F8 which contain the mean data on the worksheet. Point on the selected area's border (the mouse pointer appears as an arrow). Drag over to the chart and release the mouse button (this is known as the 'Drag and drop' method). The new (mean) data will be automatically entered into the chart.

Note: the last procedure can be undone by clicking on *Edit* and selecting *Undo Drag and Drop* from the menu.

Customizing worksheets

If you are working with a large set of data it is preferable to place data and analyses on different worksheets within a workbook in the same way that data would be kept on sheets organized within a folder. To identify the location of items within the workbook the tabs on the worksheet can be relabelled. On the worksheet with the butterfly data, click on the Sheet 1 tab once with the right mouse button. This should give you a number of options for the worksheet such as inserting, deleting, renaming, selecting or moving and copying a sheet. Select the option to rename the sheet and type butterflies on the tab, pressing the Return key to complete the renaming.

When worksheets are printed out the tab keys will not be present, so it useful to further customize the sheet so that its contents are clearly marked. From the File menu select Page Setup (Figure 3.12). A number of options appear that will allow you to change the orientation of the page and print quality, alter the page margins, insert a header and footer and the character-istics of the worksheet. We are going to add a customized header and footer, so select this tab. A number of different options may be selected to present the header and footer of each page. As an example we will customize the footer of the page (see Figure 3.13).

In the Header list box, select None and click on the Custom Footer option. Three boxes appear that represent the left, centre and right sections of the page. Start in the box on the left-hand side and type in a filename for your

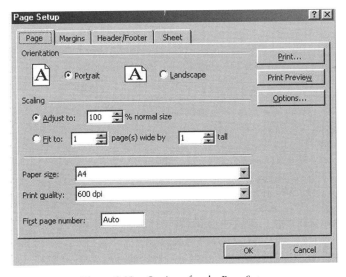

Figure 3.12   Options for the Page Setup

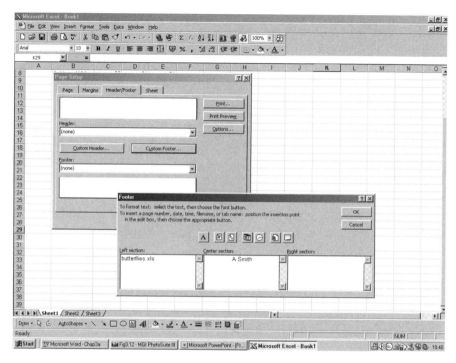

Figure 3.13   Customizing worksheets

workbook, e.g. butterflies.xls. It is always useful to have the filename recorded on a printed piece of work to make finding the file again much easier should you want to edit the information in the file. In the centre section type in your own name, particularly useful if you are submitting the printed item as coursework as the worksheet is clearly identified as belonging to you. In the box on the right-hand side we can insert (by clicking the appropriate option above) the time and date. Each time the worksheet is opened and saved, the current date and time will be recorded on the worksheet. When files are being updated by adding further information from an experiment or study, it is important to keep track of when revisions are made, so adding the date and time aids this process. Confirm the changes that you have just made (by clicking OK). On returning to your worksheet, you should be able to see the footer by clicking on the Print Preview button. You may alter the properties of the worksheet by re-entering the Page Setup menu from the top of the page. Select this option now and we will further adjust the appearance of the worksheet prior to printing. Choose the options to show the gridlines of the worksheet and to print in black and white by checking (click on the box so that an x appears) the appropriate box accordingly.

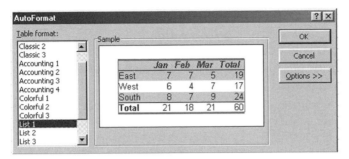

Figure 3.14   Formatting tables

Producing tables in Excel

If the information we are using in the spreadsheet needs to be used as a table in a report, the format ought to be more attractive. The butterfly data can be formatted into a table (see Figure 3.14). Click on any cell in the list of data entered on the worksheet, e.g. click on cell C5. From the *Format* menu on the toolbar, select *Autoformat*.

The range of values around cell C5 will automatically be selected and can be arranged in one of several pre-set formats. From the *Table Format* box, scroll down the options and choose a pre-set format (e.g. Simple 1) and click OK. The data are now displayed as a table in the format selected; the format may be revised as many times as you like until a satisfactory choice is made. Once you have completed your revisions, use the Print Preview option again to check that all of the items on the worksheet are going to appear in the correct position on the page, altering them if necessary by selecting and moving them. This cannot be done while in Print Preview mode; you will need to return to the worksheet to do this. After Print Preview has been used you should notice that the limits of the page can been seen as dotted lines on the worksheet. Items may then be moved around by selecting and dragging to format the worksheet within these borders, ready for printing. Having checked the worksheet thoroughly (including spellings using Tools:Spelling option) the workbook can be saved and printed.

3.2 Presenting graphs and charts

Having worked through the previous section you should now realize how simple it is to produce graphs in Excel. What is more skilful, however, is to decide the best plot for the type of data being presented. In whatever branch of science we are involved, observations are made during which we gather data.

Data can be in various forms; it can be qualitative or quantitative. Qualitative data tends to be descriptive, such as whether an individual is male or female; alive or dead; blonde, brunette, grey, etc. Quantitative data is numerical and measured with precision; for example, an individual may have a height of 173 cm and body mass of 72.3 kg.

During the course of experiments we generally collect information from our investigations (raw data) and apply the following three processes:

- *organize* the data; e.g. sort into groups, set into a table.

- *illustrate* the data in order to interpret the information from the investigation; i.e. make into a bar chart, line graph, pie chart.

- *analyse* the data using an appropriate statistical method; from the statistical test a conclusion may be drawn about the investigation.

We will be thinking about the statistical analysis of data in later sections, but for now we will look at different types of data and see how it should be presented.

Graphs and charts

Drawing a graph in Excel is easy, but does the finished item look right? Is it presented as it should be? Unless you choose the correct type of plot, producing a graph can go very badly wrong and data can be misrepresented under these circumstances. Let us begin by looking at a simple absorption spectrum.

Exercise 3.2

In a laboratory experiment the absorption for phenolphthalein, a pink-coloured indicator, was determined at a range of wavelengths. Owing to its colouration the optimum absorbance is likely to lie somewhere in the region of 540 and 560 nm, so more frequent measurements were taken within this range, although the full range of wavelengths investigated was from 450 to 650 nm. Table 3.2 shows the data obtained from the experiment. Firstly we must decide what type of graph is appropriate to draw. Where data show a trend with one item of information being related to the next in a series, i.e. they

Table 3.2   Absorption spectrum for phenolphthalein between 450 and 650 nm

Wavelength (nm)	450	500	520	530	540	550	555	560	570	580	590	600	650
Absorbance	0.2	0.51	0.60	0.65	0.68	0.72	0.73	0.73	0.67	0.63	0.59	0.49	0.31

appear in a specific order, then a line graph will show the relationship between points.

Enter the data onto a worksheet in Excel and then using Chart Wizard select the option for a Line plot (see Figure 3.15). Several types of line plots are shown for you to select the most appropriate. Choose the one described as Line with Markers at each Data Value (i.e. the one showing points and lines). Once

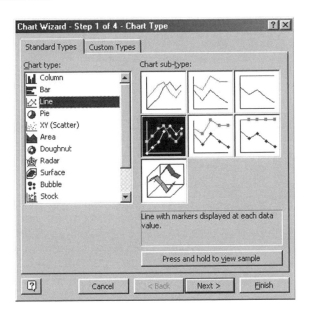

Figure 3.15   Selecting line plots

this is selected you will see that the plot is displayed for the data, but wavelength is incorrectly plotted on the graph instead of being used as the scale on the x-axis (see Figure 3.16). This is easily amended by clicking on the Series tab and, using the Remove button, highlight the wavelength label to delete this

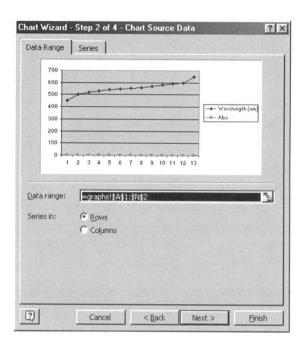

Figure 3.16   Previewing line plots

data. Then click in the Category x-labels box, select the
wavelength values and you will then see the graph being
replotted with the wavelengths on the x-axis (see Figure 3.17).
   At this point look carefully at the graph – can you see
anything wrong with the plot? (See Figure 3.18.) If you can't,
try looking at the x-axis and then think carefully about how you
would plot these data if you were doing the graph by hand. You
should then notice that the scale is not linear as it should be.
Excel has plotted the data as though each wavelength reading
is equally spaced apart, which is clearly incorrect. So what
action must be taken to remedy this plot? Use the Back button
to take the steps back to the screen where you were able to
select the type of chart that you wanted. Now click on the X:Y
Scatter option as seen in Figure 3.15. This should display the
chart where the scale on the x-axis is now equally divided (see
Figure 3.19).
   Continue through to the next step and add a title and labels
to the graph. Once this has been completed, move through to
the next step and then Finish. The graph shown in

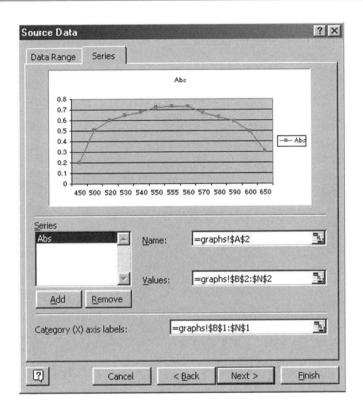

Figure 3.17 Changing data automatically plotted on graphs

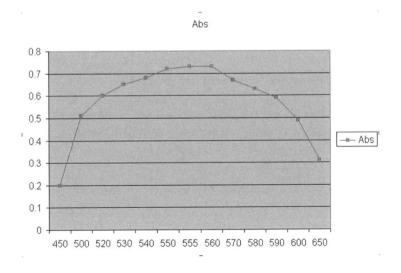

Figure 3.18 Completed graph inserted into the worksheet

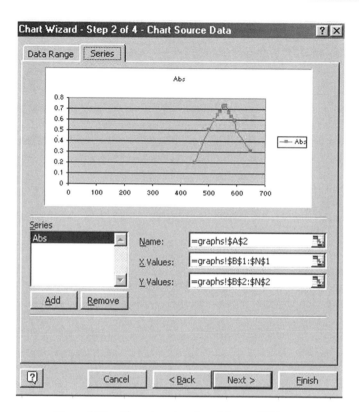

Figure 3.19   Changing plots from Line to XY Scatter

Figure 3.20 should now be what you see on your screen, but although the plot is correct, there are still a few problems. The major problem still lies with the arrangement of the scale on the x-axis. If this was plotted by hand then the axis would not begin at 0 and end at 800 nm, so this needs to be narrowed to prevent the data points being lost in the plot window as at present. This means that it is necessary to edit the graph, something that is done very easily in Excel, as just clicking on the graph will automatically take you into Edit mode. The important thing to remember here is to click at the point on the graph that you want to edit which can be any item including lines, points, background, border and gridlines. As we want to edit the x-axis, click on the graph near the x-axis and then click once with the right mouse button. As long as you haven't selected the entire plot area then you should now see an option to format the x-axis.

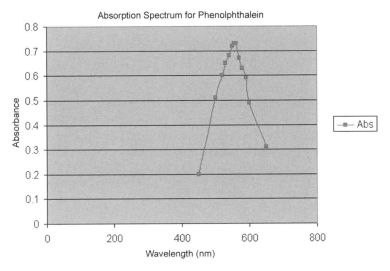

Figure 3.20   XY Scatter plot inserted into the worksheet

From the menu, click on the tab for Scale and then change the default values for the minimum and maximum values to 400 and 700, with major and minor units of 50 and 10 respectively (to determine the increments of the *x*-axis scale). The axis should now be adjusted as shown in Figure 3.21.

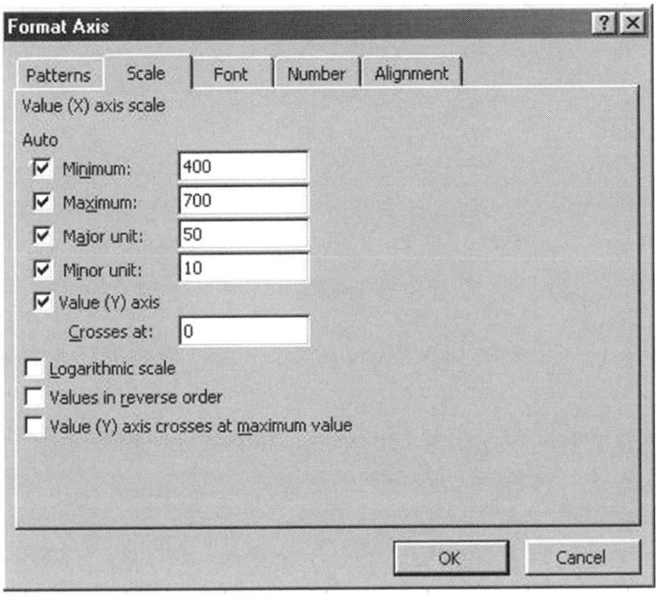

Figure 3.21   Changing the scale of the *x*-axis

The plot is now presented as it should be, but there is duplicated information on the graph. We have labelled the *y*-axis as absorbance, but as part of the automatic processes used in producing graphs Excel has provided a legend (Abs) on the plot. If we needed to have a legend, then it certainly shouldn't be in this abbreviated form; to remove the legend select it on the graph, click with the right mouse button and choose Clear from the options.

Editing plots in Excel

The way that we present our data is a significant aspect of scientific reporting and the importance of taking time to experiment with different styles in which to portray our data should not be undervalued.

Remember: charts should be presented in order to encourage the reader to make comparisons and then analyse them. The designer of the chart should ensure that the data are presented in a clear and unambiguous manner so as not to mislead or bias the reader.

Creating a good chart usually means ensuring it is as simple and clear as possible, so that its message is immediately apparent. Excel allows you to readily access a range of different chart styles; the problem is deciding which one to choose for presenting your information.

Bar charts

Bar charts are the simplest form of chart. They can be used to show numbers, proportions or ratios. In Excel, the bar charts are available as bars or columns. For the Column chart all of the bars are positioned vertically and for the Bar chart, bars are positioned horizontally across the plot. In Exercise 3.3 we will explore how to use bar charts effectively to present data where we are comparing one or more variables in an experiment.

Exercise 3.3

Copy the information in Table 3.3 onto your spreadsheet. The data presents the mean weight loss (in kg) of human subjects who volunteered for a trial in which they were asked to follow

Table 3.3   Mean weight loss (kg) during six months of human subjects on three different dietary regimes: A, B and C

Diet	Males	Females	SD (M)	SD (F)
A	11	9	2.3	1.9
B	21	18	3.6	3.1
C	13	14	2.0	1.9

three different diets (A, B and C), together with the standard deviations (SD). The data may be compared between male and female subjects. You should now be familiar enough with the Chart Wizard function to plot a column chart of this information; if you are unsure of what to do then refer back to the instructions for producing the column chart for the butterfly data in section 3.1. Clearly, in this exercise we want to be able to compare the weight loss produced by each of the three diets, but we also want to compare the comparative weight loss by each of the two sexes. In our bar chart we need to select an option where we are able to make a side by side comparison of males and females on each diet. Select the data for the males and females, including the labels (but excluding the standard deviation data) and choose Clustered Columns from the Chart Wizard options. Label and title the graph appropriately. You should then have a plot similar to that shown in Figure 3.22.

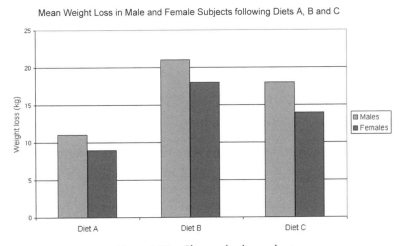

Figure 3.22   Clustered column chart

Displaying values on charts and graphs

Sometimes it is helpful to show the numerical values for plotted variables on the graph. To insert values for each column in the plot, click on the chart to enter edit mode. Select one of the bars in the chart representing data for the male subjects (place the mouse pointer over the bar and click), then right click with the mouse button. From the menu choose Format Data Series and then select the Data Labels tab. You then have the option to show the value or show the label. Choosing Show Value will result in the value being displayed over all of the bars in the series that you have selected. The process will need to be repeated for the bars representing female subjects.

Error bars

In scientific data where a mean value has been calculated, we frequently want to show the variability of the data by inserting error bars to represent standard deviation or standard error (the standard deviation and standard error are discussed in Section 4). In a bar or column chart the bars need to be inserted at the top of the bar as shown in Figure 3.23.

Enter edit mode by clicking on the chart and go to Format Data Series as described above to show the numerical values. Select the tab for Y Error Bars.

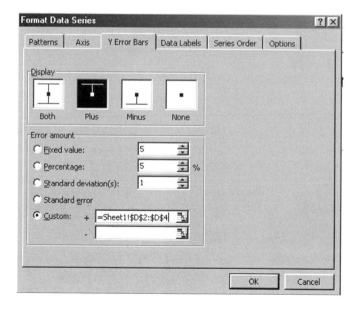

Figure 3.23   Inserting error bars

Different options are shown for the types of error bars available. Clearly for a bar series we need the Plus type, so click on this option. Below this you are prompted to indicate how the error value should be derived. To input a SD or SE based on data on the worksheet, go to the Custom box. In the + box, type in the cell references for the cells containing the standard deviations for the bars that you have selected (either the male or the female subjects). Confirm the selection by clicking OK. The chart should now be updated with error bars, but the process will have to be repeated for the second data series.

Finishing touches

It is important to ensure that all graphs are appropriately labelled and titled. Here are some hints and tips to follow to make sure that plots are properly presented.

1. Axes should have labels that explain the variables and give the units of measurement normally by providing the standard scientific abbreviation for the units in brackets, so 'kilogm' for kilograms would be unacceptable, but 'kg' is correct. Note that some units are arbitrary, as in Exercise 3.2 where absorbance of phenolphthalein was measured; absorbance does not have units.

2. Symbols for units in graphs can be inserted into plots using a code for each symbol. Quite frequently we need to insert symbols such as °C for degrees Celsius or μg for micrograms. Symbols have unique codes that can be entered by pressing the ALT key and the appropriate numeric code on the number pad (N.B. it must be using the number pad otherwise this feature will not work). For example μ will be inserted by pressing ALT then 0181. A list of useful codes is provided in the Appendix.

3. Titles of graphs should be short and concise. The chart title should be clear about:

   - The group or sample that is being described (subjects, male and female).

   - The variables involved (diet A, B, and C).

   - The type of data presented (weight loss (kg)).

In scientific investigations we frequently perform experiments or trials and there is a tendency to start titles by writing 'An experiment to show . . .' or 'An investigation of . . .'. It is not necessary to start a title in this fashion and it is a practice that should be avoided. Titles are more readable if the important information is placed first, for example,

'Mean weight loss in male and female subjects
following diets A, B and C (mean+SD)'.

as opposed to:

'An investigation of the mean (+SD) weight loss in male and female
subjects following three different dietary regimes (A, B and C)'.

Clearly the first title is succinct and provides the message of the graph very clearly.

Framing and gridlines

Most charts benefit from having a frame, especially if they contain gridlines. Excel has the facility to change the background of a chart to different colours. A light shading is preferred so that it does not interfere with the emphasis of the chart. Some caution needs to be used when printing in black and white as even what appears as a pale grey background can spoil the appearance of the chart, particularly where a contrast is needed between the columns on the chart and the background colour.

Gridlines make most charts easier to read as they provide references for us to judge the values of bars or points. In very simple charts, however, they can prove to be a distraction and are better removed. The formatting of gridlines is also important as they need to be as unobtrusive as possible. They should be spaced at appropriate intervals. Too many will make the chart appear busy; whereas too few will not provide sufficient reference points.

Formatting gridlines

Click on a gridline, then right click the mouse button. Select options under Style and Weight to reformat the gridlines.

Setting the correct proportions for the chart

Sometimes when a chart is placed directly onto the worksheet it appears as a small and narrow plot, as shown in Figure 3.24. You should ensure that the

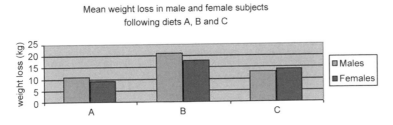

Figure 3.24   Example of a graph that needs re-sizing

finished item does not look like a widescreen TV, by re-proportioning the graph and adjusting the size of the font for titles and labels when necessary. There is sometimes a tendency to accept whatever the software produces without questioning whether the message is clearly conveyed in the graph or whether extra work needs to be done to adjust some of the components of the plot. The editing features of Excel are very easy to use and there is no excuse for presenting substandard plots and blaming it on the software.

Exploring different types of bar charts

Stacked column bar charts

This is a slightly different way in which we are able to convey information and the message of the graph might be different from the graph where bars are positioned side by side. It is useful for summarizing what has occurred in an investigation, but allows us to see data from subgroups and therefore assess the contribution of components within the experiment. The mean data for the diets could be represented in this way if, for instance, we were interested in comparing the total mean weight loss between male and female subjects, but wanted to see the individual contribution of each diet to the weight loss as a whole. Using the Chart Wizard function, re-plot the data selecting the Column chart option with stacked columns. As you produce the graph you must ensure on Step 2 that you change the Series option so that the data are organized in rows instead of columns. This will produce the side by side comparison of the mean weight loss for males and females. Your graph should appear as in Figure 3.25. We can see from this graph that the mean weight loss was greater in the male than in the female subjects; whereas in our previous graph, Figure 3.22, the emphasis was on comparing the weight loss between diets and showing that Diet B produced the largest weight loss. If you have plotted the graph and it appears exactly as in Figure 3.25 you may want to consider how the appearance of the plot could be made clearer. The legend, for instance, shows Diet A,

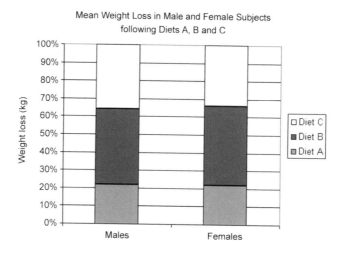

Figure 3.25   Stacked column chart

Diet B and Diet C as these are the labels on the worksheet. Sometimes we may want to change these or correct mistakes. Enter edit mode by selecting the graph and choose Source Data after clicking with the right mouse button. This takes you back to the step where you selected rows instead of columns. Click on the Series tab. From here you are able to select the data in each row and in the name box rename each label in the legend, as shown in Figure 3.26.

A further option that we might want to include is to show a table containing the data itself beneath the plot. Once the chart is complete you can edit the graph (by clicking on it) and then select Chart Options from the menu. Click on the Data Table tab and by selecting Show Data Table, as can be seen in Figure 3.27, a data table is displayed beneath the plot.

Grouped 3-D bar charts

The information can be conveyed again slightly differently by using a three-dimensional bar chart. Here bars may be placed in front of or behind each other and so give emphasis to components of the plot.

In Excel, re-plot the weight loss information; this time select the 3-D Column option, placing the data back into columns instead of rows. On the three-dimensional plot it would be more appropriate to have the label 'weight loss' at the top of the axis with the text written horizontally (remember we read from left to right) rather than written vertically so that the reader needs to turn through 90° to be able to read it. To adjust the position of the label, select the

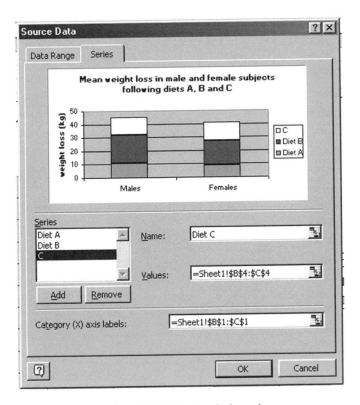

Figure 3.26   Editing the legend

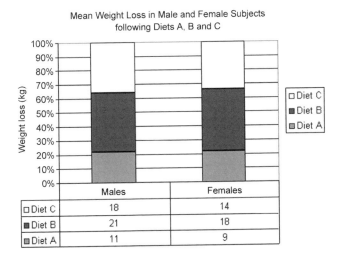

Figure 3.27   Displaying data beneath a graph

label to enter edit mode and right click the mouse button. From the options choose Format Axis Title. You are then confronted with options to alter the text alignment of the label. Use the mouse button to alter the orientation of the text as shown in Figure 3.28 and confirm your choice. You should then have a plot that looks very similar to that shown in Figure 3.29.

Finally, experiment with these data by changing the overlap of the bars in the chart. If we look at the plot we can see that the data for the male subjects 'overshadows' that for the females. This is because the weight loss was greater

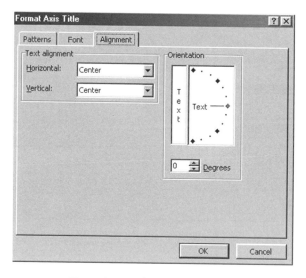

Figure 3.28   Aligning text in titles

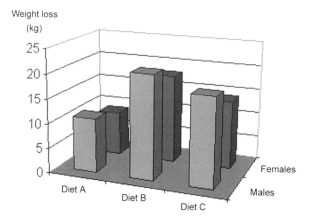

Figure 3.29   Three-dimensional plots

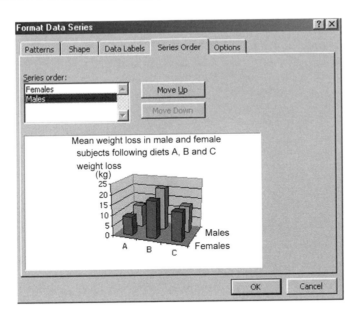

Figure 3.30   Changing the series order in graphs

for the male subjects. It would be preferable for the female data to appear at the front of the chart, so how do we accomplish this? By clicking on the plot, enter edit mode. Select the bars representing the male subjects and click on the right mouse button. Choose Format Data Series from the menu and then select the Series Order tab. Using the move up and move down buttons you are able to alter the position of the bars on the graph as shown in Figure 3.30. The data are more aptly presented with the female data columns being in front of those for the males, so select this option and return to the worksheet. Although the display is improved, to provide further contrast between the male and female data it would be better if the bars at the front of the graph were lighter than those behind. Using the editing options that you applied in section 3.1, change the bar colours until you have a plot similar to that shown in Figure 3.31.

Printing bar and column graphs

The graphs shown on your computer monitor are usually impressive as some good comparisons are shown using appropriately contrasting colours. When it comes to printing, however, some of the contrasts may be lost, particularly where very light colours have been used against an equally light background. For bar, column and pie charts it may be necessary to select patterns. Here are

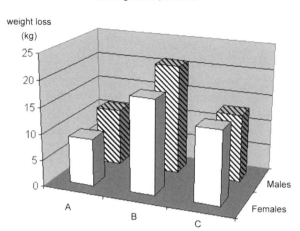

Figure 3.31   Completed graph showing emphasis by changing colours and patterns

a few tips on how to make the patterns on your plots look equally as good when printed in black and white:

- Use dots and lines as filler patterns as these give good results. Lines are better if they are slanted rather than horizontal or vertical. Avoid some of the graduated shading that is available in Excel as this may cause problems in contrasting with the background shades.

- Avoid using patterns that are too busy. These detract from the plot.

- Columns that are completely black should be avoided as these may smudge on printing and may also dominate the chart. They also use up vast quantities of ink or toner if you are producing full-page plots.

- White may be used for emphasis and does not have the effect of being too overpowering. It is particularly effective if you are trying to emphasize a 'control' or 'no response' group.

- Avoid using too many different patterns on a chart as the result is confusing. Do not place patterns that are similar too close together otherwise the contrast is lost.

Pie charts

These are the main alternative to bar charts and are useful in making comparisons of proportions. Using a pie chart it is difficult to read individual values, particularly where there are several categories, so the pie chart tends to be used for the purpose of providing an overview. By using the feature in Excel to remove a 'slice' of the pie, a particular aspect of the data can be emphasized.

Taking the data from Exercise 3.3 (Table 3.3) we will see how to construct a pie chart to represent the decrease in body weight for the male subjects. Using the data on your worksheet, select the data for the male subjects and click on the Chart Wizard button. From the list of available options select Pie with a 3-D visual effect. Continue through the chart options to complete the plot which should be similar to that in Figure 3.32. Although the three-dimensional pie is effective it would be easier to judge the different proportions if the position of the pie was adjusted. This is accomplished in Excel by selecting the pie; to

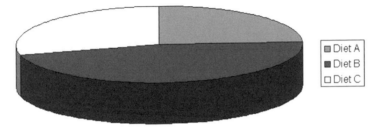

Figure 3.32   Pie chart

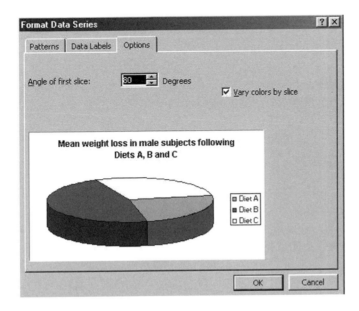

Figure 3.33   Changing the angle of the first slice

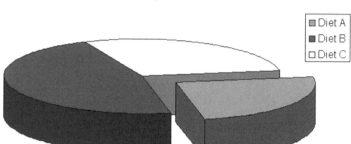

Mean weight loss in male subjects following Diets A, B and C

Legend: Diet A, Diet B, Diet C

Figure 3.34   Pie chart with slice removed

accomplish this click on it, but in doing so make sure that handles appear on every slice of the pie. Finding exactly the right selection can sometimes be difficult; editing with different selections can pull apart slices or expand the top or sides of the pie. You will need to experiment with these features to find out exactly how they work. Once you have successfully selected the pie pieces, however, you should then be able to select the option to Format Data Series. From this go to the Options menu. Here you are able to move the angle of the first slice. By increasing the angle you will cause the pie to rotate, as shown in Figure 3.33. Try this option until you reach the point where you feel that the pie pieces are now much easier to compare than in the original plot, and then confirm your choice. The plot shows that the least weight loss was experienced with Diet A. To place emphasis on this point we could remove or 'explode' a piece of pie. By clicking on the slice of pie for Diet A it should be possible to select and then drag the slice from the other pieces. Try this for yourself. The finished plot should be comparable to that shown in Figure 3.34.

Line graphs

Line graphs are used to compare two variables and show the relationship that exists between them. Usually the independent variable is plotted on the x-axis and the variable that is dependent on x on the y-axis. An independent variable is one that is controlled by the experimenter, so this will include variables such as time, temperature, pH, etc. The dependent variable is dependent on the value of x and so will change with x. Line graphs show an ordered relationship between sets of data so that if the value of one variable is known the graph may be used to predict the value of the other.

Table 3.4   Concentrations of drugs A and B against time

Time (h)	Concn drug A (µg/ml)	Concn drug B (µg/ml)	SD (A)	SD (B)
1	100.1	120.2	5.6	6.6
2	50.2	100.3	2.1	5.4
3	25.5	80.4	1.9	4.3
4	20.2	62.5	1.4	3.6
5	15.6	51.4	1.1	2.0
6	12.1	39.6	0.8	1.5
7	10.3	33.5	0.5	0.9

We will use as an example a kinetics plot where the concentration of a drug is seen to change with time. In the example in Table 3.4 there are two drug concentrations that are being investigated so we can use a multi-line graph.

Exercise 3.4

Enter the data from Table 3.4 on your worksheet. Using the option for XY (Scatter) and Data points connected by smooth lines, plot a multi-line graph for both drugs on the same plot. In producing the labels for this plot you will need to insert the units for concentration. These are $\mu g \cdot ml^{-1}$. To insert symbols into Excel that will appear on worksheets and in graphs and charts you can use the symbol codes (listed in the Appendix). To insert a symbol press the Alt key on the computer, then enter the numerical code using the Number pad on the right-hand side of the keyboard. On releasing the Alt key, the symbol will appear on your worksheet. Complete the plot by adding titles and labels. You should now be familiar with inserting error bars, so include the standard deviation on your plot, placing + error bars on the upper line and − error bars on the lower line. Your graph should appear as in Figure 3.35.

We will now see how we can transform the data by using a semi-logarithmic plot. These plots are often used with kinetic data where the y-axis is represented logarithmically. Click on the chart to enter edit mode and select Chart Type from the edit menu (produced by right clicking the mouse button). Click on

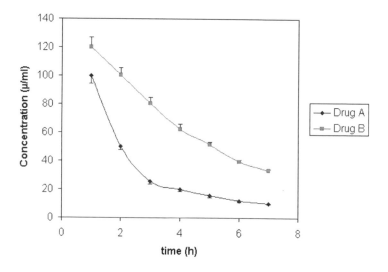

Figure 3.35   Line graphs

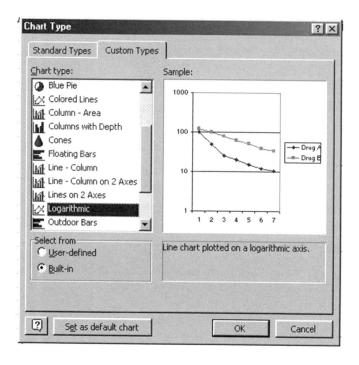

Figure 3.36   Selecting a logarithmic line graph from chart options

the Custom tab. Here you will find a number of graphs that do not appear under the standard options. Select Logarithmic from the list. A preview of the graph appears on which the *y*-axis is a logarithmic scale as seen in Figure 3.36. Confirm your choice and complete the graph.

Combination plots

Sometimes we may want to demonstrate a change in two variables, each with different units of measurement, on the same graph. This is where we need to use what is known as a combination plot. This plot has two *y*-axes; different units and scales can be used on each axis and the data are presented as a combination of a bar chart and line plot.

Exercise 3.5

The data in Table 3.5 compares the change in heart rate and diastolic blood pressure in a hypertensive patient during a period of moderate exercise on a treadmill. As we are interested in how each variable changes with time a combination plot would be ideal to show how the two variables might be related.

   Enter the data from Table 3.5 on your worksheet and using Chart Wizard, choose one of the combination chart options. This will again be found on the Custom Types selection under

Table 3.5   Mean diastolic blood pressure in a hypertensive patient during moderate exercise

Time (minutes)	Diastolic BP (mmHg)	Heart rate (bpm)
10	80	80
20	85	85
30	93	90
40	98	100
50	99	110
60	105	120

Line-Column on 2 axes (we select two axes rather than one as the units for blood pressure and heart rate are different). The preview for the chart will show that all three variables – time, blood pressure and heart rate – are plotted on the graph. Clearly this is wrong as the x-axis should be time (as this is the independent variable) and not the arbitrary numbers inserted by Excel. To amend the plot select the Series tab and click on Time from the Series list and then Remove. The time data now need to be re-inserted under category x-axis labels (as shown in Figure 3.37), so click in this box and insert the cell references for time, but excluding the label Time from your selection. The preview shows the graph correctly plotted and we can complete the graph by adding titles and then make our comparison of the change in heart rate and blood pressure in the patient over time.

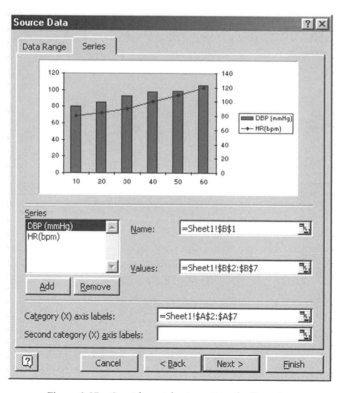

Figure 3.37   Semi-logarithmic options for line plots

WEB SUPPORT – SECTION 3

Here you will find plenty of data with which to experiment with different types of plots. You will be able to compare the finished result with ready-prepared charts so that you can see whether you have presented the data correctly. You'll also find more hints and tips on data presentation, plus any information about Excel updates that affect plotting functions.

4

Preliminary Data Analysis

Having reviewed data from investigations by plotting graphs, we may conduct some preliminary statistics before moving on to testing the data. Usually we are interested in looking for trends in data, determining the variability of results and considering its validity as a representative sample from the population from which it was drawn. This section reviews some of the techniques used for preliminary data analysis.

4.1 Descriptive statistics

As the name suggests, these are statistics that we calculate in order to summarize the data from our studies. They are used to give a description of the data by determining measures of location and to express its variability. Each of these aspects will be discussed in turn.

Measures of location

There are three main types of measures of location, these are known as the (arithmetic) **mean**, also known as the average, the **median** and the **mode**. Each has different properties and uses.

Data Analysis and Presentation Skills by Jackie Willis.
© 2004 John Wiley & Sons, Ltd   ISBN 0470852739 (cased)   ISBN 0470852747 (paperback)

Table 4.1   Number of hours per week spent watching television by a group of students

									Mean	Mode	Median	SD
12	13.5	10	10.5	7	10.5	12	9.5	10.5	10.6	10.5	10.5	1.8

The mode

The mode is the category or class of variable with the most observations in it, i.e. the most frequently occurring value. Table 4.1 shows the number of hours a sample of students spent watching the television each week. As we can see from the table, the mode is 10.5 hours as this is the most frequently occurring time. Sometimes there may be two values for the mode, in which case the sample is said to be bimodal. The mode does not indicate the centre of the sample, only those values that occur the most often.

The mode is very easily calculated in Excel. Enter the raw data from the worksheet (the raw data is the individual values for each of the students and so will exclude the summary statistics that have been calculated). Choose a cell on the worksheet in which you would like the modal value placed, then click on the Paste Function (see Section 3.1) and select MODE from the Statistical menu. You will be prompted to enter the cell references for the cells that contain the raw data, confirm your selection and the value for the mode should appear in the cell that you selected on the worksheet.

The median

If all of the observations in a set were placed in ascending order, then the median would be the middle observation. The median will have as many observations above it as below it. If we look again at Table 4.1, but this time sort the values in ascending order, we can see that 10.5 hours is the middle value as there are exactly four values above and four values below this number.

									Mean	Mode	Median	SD
7	9.5	10	10.5	**10.5**	10.5	12	12	13.5	10.6	10.5	10.5	1.8

The median therefore gives us an indication of the value in the central location of the sample, but it does not summarize all of the data. The median provides the middle value of the distribution. Where there is an even number of values, the median will be the average of the two middle values (e.g. if there were eight in our sample and the two middle values were 10 and 10.5, then the median would be 10.25 hours).

The median can be calculated from Excel in the same way as the mode. Using the data entered on the worksheet, click on the Paste Function and select

MEDIAN from the list of Statistical functions. After entering the cell references the value for the median will appear on the worksheet.

The mean (average)

In contrast to the median, the mean summarizes all of the data and is calculated by adding all of the values and dividing the sum by the number of observations. So from the data in Table 4.1 the mean value would be $95.5/9 = 10.6$ hours. Although the mean provides a value that includes all of the data, one problem is its sensitivity to any extreme values that may occur within a data set. If we had an additional student in the sample that watched television for 40 hours per week, the mean value would become $(95.5 + 40)/10$, i.e. $135.5/10 = 13.6$ hours. Clearly the value of the mean is no longer a good measure of the centre of the sample. If we compare this with the median value, the additional observation does not alter it in any way as the median value is still 10.5 hours.

We have already used Excel to calculate the mean as this was used in Section 3 with the butterfly data. The mean is denoted as the AVERAGE in the Statistical functions in Excel.

Choosing between using the median or the mean

When deciding which measure to use, the shape of the distribution from which the sample is taken becomes the deciding factor. Where a distribution is symmetrical, showing a normal (bell-shaped) pattern as can be seen later in this section in Figure 4.4, the mean value is preferred as it uses all of the observations in its calculation. Where a distribution is skewed, therefore containing an excess of extremely large or extremely small observations, the median is preferred as it is insensitive to these extremes. If the mean were to be used, a shift in its value would have occurred either to the left or to the right of the distribution, depending on whether is it positively or negatively skewed, and therefore the mean value would be clearly inappropriate. These aspects of distributions are further discussed in section 4.2.

Measures of variation

The measures of variation of a set of observations are described by the range, variance and standard deviation. Each of these is used to determine the variability within a set of data. If we return to the data in Table 4.1 and include

Table 4.2   Number of hours per week spent watching television by two groups of students

										Mean	Mode	Median	SD
Group 1	9	10	9.5	8.5	11	9	10.5	7.5	20	10.6	9	9.5	3.7
Group 2	12	13.5	10	10.5	7	10.5	12	9.5	10.5	10.6	10.5	10.5	1.8

data from an extension of the original investigation. All of the students in the original study lived in halls of residence; we will assign them as Group 2. A further group of nine students was investigated, all of whom lived at home. The number of hours per week spent watching television was compared between the two groups. The data are shown in Table 4.2.

Simply by looking at the information in the table we can see that there is a difference between the two groups. The mean number of hours spent watching the television is exactly the same for each group, but there is clearly more variability in the number of hours in Group 1 than in Group 2, and values for the median and mode are different. There needs to be some means of representing the variability between the groups.

The range

The range is a very basic means of expressing the extent of variation in a sample; it is simply the difference between the maximum and minimum values. So for Group 1 this will be:

$$20 - 7.5 = 12.5 \text{ hours}$$

and for Group 2 this will be:

$$13.5 - 7 = 6.5 \text{ hours}$$

Like the median and mode, the range only uses a small part of the data (largest and smallest values) and so does not reflect the true variation between all of the values.

The standard deviation and variance

The standard deviation and variance indicate how closely packed around the mean the values in a variable are. The standard deviation and variance use all the information in the sample and have a number of mathematical properties which enable them to be used in various statistical tests.

Variance:

$$\text{variance} = \frac{\sum (x - \text{mean})^2}{n - 1} \qquad \text{(Equation 4.1)}$$

where x is each individual observation in the sample.

The variance is calculated by subtracting the mean from each individual value in the sample and taking the total of the sums of the squares of the deviations from the mean values. The resulting value for the variance is usually a large number in relation to the mean value. The measure of variation is therefore taken to be the standard deviation which is the square root of the variance.

Standard deviation:

$$\text{SD} = \sqrt{(\text{variance})} \qquad \text{(Equation 4.2)}$$

From looking at these equations it is easy to see how the variance and SD are able to represent the variability in the data in comparison to its mean value. The more variation there is in a sample, the greater the deviations of the values from the mean. As the standard deviation and variance use all of these deviations in their calculation, they truly reflect the variability in the data. Looking back at the data in Table 4.2 we can see how this helps us to interpret the results for the two groups of students. The mean value may be the same, but there is far more variability in the first group than in the second as the standard deviations are 3.7 and 1.8 hours respectively. The variability for Group 1 can be mainly attributed to one very large value: one student watched television for 20 hours. This clearly has an effect on the distribution of the results, as the other students tended to have much lower viewing times. This is confirmed by comparing the values for the median, mode and range.

We have already seen how Excel can be used to calculate the standard deviation of data, using the examples with the butterflies in section 3.1. A far more useful facility in Excel is to use the Descriptive Statistics function that will supply all of the descriptive statistics for a set of data and so save time in calculating each parameter individually.

Descriptive statistics in Excel

Input the data in Table 4.2 into an Excel spreadsheet. From the Tools menu select Data Analysis.

Note: If the Data Analysis option does not appear at the bottom of the Tools menu then you will need to load this function either from your network or from the Microsoft Office CD. From the Tools drop down menu, select Add-Ins and from the list provided check the box against Analysis ToolPak. After you have selected OK, the ToolPak should be loaded and you should then find Data Analysis under the Tools menu when this is reselected.

Choose Descriptive Statistics from the list provided. A dialogue box should then appear in which you input the range of cells for the data arranged on the worksheet. Include the labels in the selection and then check the box Labels to show that these are included, as shown in Figure 4.1. If your data are in rows rather than in columns then also ensure that you change the option in the dialogue box. Check the Summary Statistics to indicate that you want these displayed and then having chosen where on the worksheet the results should appear (it is usually a good idea to choose a new worksheet where there is a lot of data), click OK.

Your workbook will be updated with a table of summary statistics as shown in Figure 4.2.

Standard error

One of the descriptive statistics produced by Excel is the standard error, sometimes abbreviated as SEM (standard error of the mean). There is no function in the Paste Function to calculate this value by itself, so it has to be calculated by using a formula. The standard error is by definition an estimate of the standard deviation of the distribution of the mean, describing how spread out the distribution of the population from which the sample was taken actually is. The mean that is calculated from a sample is never the same as the value for the mean if the data for the entire population were to be included. The standard error provides an estimate of how closely the sample mean represents the true mean for the population. So when the standard error is low, it is more likely that the sample mean is a good reflection of the value for the

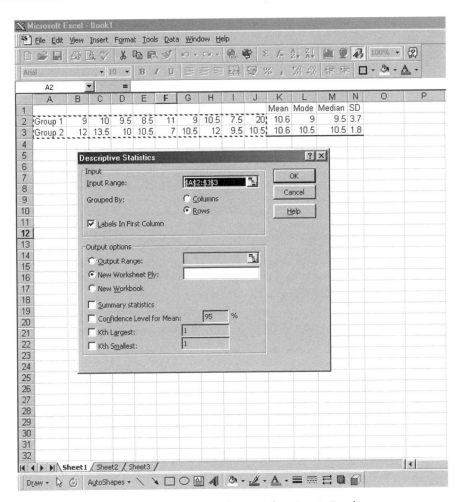

Figure 4.1 Descriptive Statistics functions in Excel

true population mean. The equation for the standard error may be seen in Equation 4.3:

$$SEM = \frac{standard\ deviation}{\sqrt{number\ in\ sample}}$$

(Equation 4.3)

If we wanted to check that the value of the standard error calculated in the Descriptive Statistics function was correct then we would insert the following formula into a cell on the spreadsheet, using the data from Group 1 as an example:

$$= 3.7/SQRT(9)$$

Group 1		Group 2	
Mean	10.55555556	Mean	10.61111
Standard Error	1.231655515	Standard Error	0.611111
Median	9.5	Median	10.5
Mode	9	Mode	10.5
Standard Deviation	3.694966546	Standard Deviation	1.833333
Sample Variance	13.65277778	Sample Variance	3.361111
Kurtosis	7.031469585	Kurtosis	1.327734
Skewness	2.537260003	Skewness	-0.50142
Range	12.5	Range	6.5
Minimum	7.5	Minimum	7
Maximum	20	Maximum	13.5
Sum	95	Sum	95.5
Count	9	Count	9

Figure 4.2   Descriptive Statistics for the television viewing data

where 3.7 is the standard deviation of the sample, for which there were nine observations, so it could be calculated by:

$$= STDEV \text{ (range of values in sample)}/SQRT \text{ (number in sample)}$$

When presenting graphs showing mean values it is usually expected that error bars are included by using either the standard deviation values to demonstrate the variability in the sample, or the standard error to demonstrate the deviation of the sample from the true population mean.

Kurtosis and skewness

Values for kurtosis and skewness are also produced by the Descriptive Statistics function. These are used to characterize the data relative to a normal distribution. Skewness is a measure of symmetry. Where data are symmetrical about the mean the skewness would be expected to have a value of around 0. If data are skewed to the left or right then the centre of the data is not around the mean and so a negative or positive value for skewness would be obtained. Skewed distributions are further discussed in section 4.2.

Kurtosis compares the shape of the data to a normal distribution and is a measure of whether the data tend to be peaked or flat. Where a high value for kurtosis is observed, data show a distinct peak about the mean and then decline rapidly. For lower kurtosis values, data are more spread out, giving a flat top to the shape of the distribution rather than a peak. A value of around 3 would represent a normal distribution.

Coefficient of variation

This function also does not appear in Excel but is a very useful parameter to calculate. The coefficient of variation represents the standard deviation as a percentage of the mean value; it is particularly useful when comparing the reproducibility of results. In quantitative analytical methods, the coefficient of variation is used as a measure of precision in quality control determinations.

The coefficient of variation is calculated as shown in Equation 4.4:

$$\text{coefficient of variation} = \frac{\text{standard deviation}}{\text{mean}} \times 100\% \qquad \text{(Equation 4.4)}$$

The coefficient of variation is usually given as a percentage and expresses the variability (from the standard deviation) of the sample compared to the mean value. It is a useful parameter to use when comparing two or more samples with different means to see if the variability is the same in each sample.

Exercise 4.1

If we take as an example a laboratory analysis conducted by two students. Each performed an assay to determine the protein concentration of a sample containing $125\,\mu g \cdot ml^{-1}$ of protein. Each repeated the analysis 10 times and the results are shown in Table 4.3.

Enter the data on a spreadsheet in Excel and perform the descriptive statistics on the data. Using the data for the mean and standard deviation for each sample, enter the following equation into one of cells on the worksheet, inserting the appropriate value for the mean and standard deviation in each case:

= (value for standard deviation/value for the mean)*100

When comparing the means you should find that both students have a mean value of $125\,\mu g \cdot ml^{-1}$ from their protein determinations, but student 2 has a more precise technique as the coefficient of variation is 2.3 per cent for their analysis compared with 7.3 per cent for student 1.

Table 4.3   Protein determinations performed by two students with a sample 125 μg·ml^{-1}

Student 1	125	120	122	130	115	140	130	121	125
Student 2	121	124	127	122	125	126	128	126	126

4.2 Frequency distributions

When we conduct scientific investigations, we collect data by taking samples from much larger populations. In order to learn something about the population we use descriptive statistics, but we also need to examine the characteristics of the distribution in order to determine the best way to summarize and analyse data.

In Section 3 we learnt about presenting data in the form of bar charts. We can draw bar charts of data in which we measure frequency (the number of times a particular occurrence takes place, for example the number of individuals in a population with blue eyes); if we draw a line at the midpoint of the bar then we obtain a frequency polygon. Increasing the number of bars in the plot, providing there is sufficient data to do so, will eventually produce a smooth curve, the shape of which will tell us something about the characteristics of the population. Figure 4.3 shows how a frequency polygon may be produced from a bar chart, using data showing height of a sample of adults from a population. This type of bar chart is known as a **histogram**.

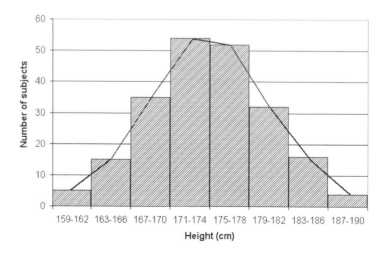

Figure 4.3   Normal distribution of heights of subjects

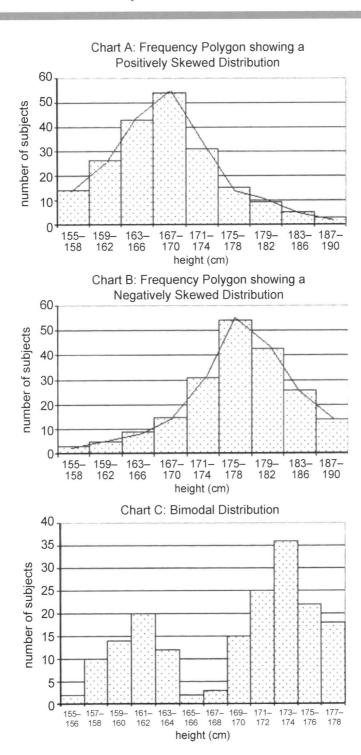

Figure 4.4   Skewed and bimodal distributions

Where the resulting frequency polygon resembles a bell-shape we can see that the population is symmetrical and the shape of the curve is said to be 'bell-shaped'. At each end, or tail, of the curve, there is a small number of extremely small or extremely large values, but the majority of the observations fall in the middle part of the curve, i.e. they are centred around the value for the mode. If we were to calculate the mean and the median for these data we would find that values would be virtually identical. A curve is said to follow a normal distribution where this occurs, so as the mean will reflect the central tendency of the distribution it should also resemble the midpoint of the distribution, represented by the median.

It is useful when considering the shape of a population to look at the tail of the curve that is produced. In Figure 4.4 we can see two distributions that cannot be normal as they do not follow a bell-shape; these are known as skewed distributions, of which there are two types, positive and negative (see also the subsection 'Descriptive statistics in Excel' in section 4.1).

A distribution with a positive skew will contain more extremely large values than extremely small ones and therefore resembles Chart A. Clearly the mean calculated for these data would not represent the central location of the distribution. Similarly, if we consider Chart B there are clearly more extremely small values than extremely large ones, in which case the data are negatively skewed. For each of these curves, the best measure of the central tendency for the data would be represented by the median value and not the mean.

Sometimes the shape of this distribution appears as if two normal (bell-shaped) distributions have been combined together, as shown in Chart C in Figure 4.4. This would suggest that there is a mixed population, which might arise where a population contains two species.

In plotting these curves we have split the data into groups, or intervals, that are equally spaced apart. The more intervals we are able to divide the data into, the more well-defined the curve becomes. We will see how by using raw data for heights of individuals we are able to produce a frequency distribution and how the Excel Paste Function may be applied to aid this process.

Exercise 4.2

The data in Table 4.4 have been collected from a sample of 40 individuals from a population. Enter the data in one column in a new workbook in Excel. The height of each subject was recorded to the nearest centimetre, so in terms of the absolute accuracy of the results, a person whose height is between

Table 4.5 Height (cm) of forty individuals from a university tutorial group

147	154	157	163	163	165	168	171	173	177
151	155	152	161	161	169	169	172	175	177
158	155	159	161	164	167	165	182	175	172
154	156	165	162	160	188	176	173	170	167

153.5 and 154.4 cm would still be recorded as 154 cm (by rounding up or down). Height would therefore be described as being a **continuous variable**, but because we are taking recorded measurements correct to the nearest centimetre, we are sampling **discrete** values.

The data on the worksheet make little sense as they stand and need to be organized. The first, most obvious step is to place them in order. Using the Data|Sort command (as described in Section 3), organize the data into ascending order. Look down the column of data to see the results. We can now see that the smallest (minimum) value for height is 147 cm whereas the largest (maximum) is 188 cm, so the heights of the individuals range from 147 to 188 cm. Even after sorting, the data are still difficult to interpret as each value has to be examined in relation to all the others (and what if we had thousands of measurements?). The next stage is clearly to group the data; this is done by dividing it into **classes** – with evenly spaced **intervals** between groups.

Rule: When data are divided into intervals it should usually be into no more than 10 intervals and no less than five intervals. Each interval should be of an equal width.

To determine how many groups to divide the data into, count the number of observations. In this case $n = 40$.

Take the square root of the total and round to the nearest whole number ($\sqrt{40} = 6.325$), i.e. 6.

Excel is able to automatically group frequency data but needs to be given the parameters by which to do this. You will first of all have to make some decisions about your data.

Firstly, look at the range of the data (147–188 cm). In order to group the data we need to work out how to have evenly spaced intervals. Clearly, if we group the data into six classes then the interval between them should be:

$$\text{interval} = \frac{(\text{highest number} - \text{lowest number})}{\text{number of classes}} \qquad \text{(Equation 4.5)}$$

$= (188 - 147)/6$ which gives us an answer of 6.83, so the interval between the classes should be 7 cm. In Table 4.5 we can see how the data need to be grouped. The number in the class column is the lower value for the class and moves upwards in steps of 7 cm.

The first class (147–153) will contain the discrete values:

$$147 \quad 148 \quad 149 \quad 150 \quad 151 \quad 152 \quad 153$$

where 147 is the **lower class boundary** and 153 is the **upper class boundary**.

In Excel, data are divided into bins (classes) in which you define the upper class boundary. Using these bins, frequency data can be produced from a list of observations, so you will need to enter onto your data sheet the classes (bins) in which you want to categorize your data. On the worksheet, type in the upper class boundaries for the data (so from Table 4.5 the upper class boundaries will be 153, 160, 167, 174, 181 and 188; enter the data in one column).

Table 4.5   Classes for the student height data

Height (cm)
147–153
154–160
161–167
168–174
175–181
182–188

Using the histogram function

From the Tools menu select Data Analysis and from the list provided choose Histogram. A dialogue box should appear as shown in Figure 4.5. Enter the input range of the data and then the range of cells containing your bins. Click on the Chart Output box so that a histogram of the data is plotted on the worksheet and confirm your selections.

A table should now appear on the worksheet in which the data has been placed into the six classes provided. The data should be presented as in Table 4.6.

We now have what is known as a frequency distribution of our data. The data is also presented in a histogram as in

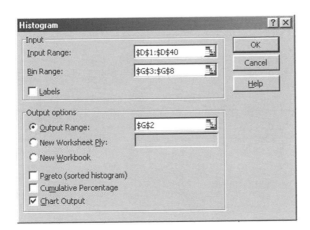

Figure 4.5   Using the Histogram function in Excel

Table 4.6   Output table from Excel showing grouping of data into bins

Bin	Frequency
153	3
160	9
167	12
174	9
181	5
188	2
More	0

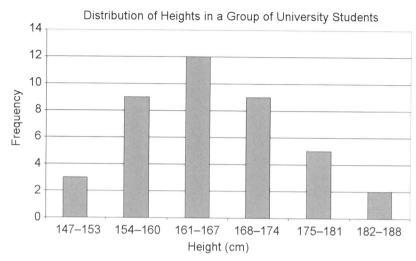

Figure 4.6   Frequency histogram for heights of university students

Figure 4.6. We can see that this appears to approximate to a normal distribution, but it is difficult to be certain with a limited number in the data set. If the sample were larger we could increase the number of bars in the frequency histogram by setting classes (bins) closer together; the histogram would appear more as a smooth curve. The shape of the distribution is represented by the shape of this curve.

When considering the statistical testing of data, it is important to establish in conducting an experiment:

(a) whether a sample is sufficiently large enough to represent the population as a whole.

(b) that the characteristics of the population are known (i.e. normal, skewed, bimodal) in order to choose the correct test to be applied to the data and the most appropriate summary statistics to describe it.

4.3 Correlation and linear regression

Sometimes we conduct an investigation to determine whether there is an association between two variables of interest. The starting point of finding out whether such a relationship exists is by visually examining the data in the form of a scattergraph; this will show us whether:

- there is a distinctive trend between the two variables (x and y) or the relationship is entirely random, i.e. related or independent
- the relationship, where found, is rectilinear or curvilinear
- the relationship is positive or negative

We can then explore associations statistically by quantifying the correlation between variables; the closeness of the relationship is expressed by the correlation coefficient, r.

When $r = +1$ the two variables are positively related.

When $r = -1$ the two variables are negatively related.

A value of 1 for r indicates an undisputed relationship between x and y, so this would indicate a perfect correlation between the two variables. A value of 0 would indicate no possible relationship between x and y, so there would be no

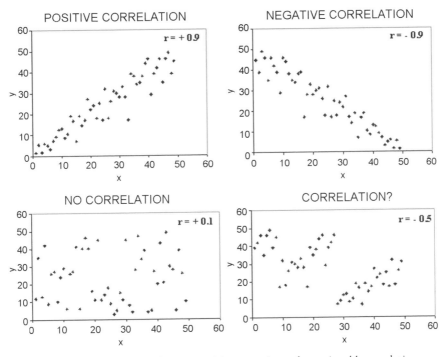

Figure 4.7   Scattergraphs showing positive, negative and questionable correlations

correlation whatsoever. In practice these values represent two extremes and most correlation coefficients lie in between these values; a judgement on the association between variables is therefore made on the proximity of the values to either 0 or 1. Figure 4.7 shows a number of scattergraphs and their corresponding correlation coefficients.

Correlation

In order to determine statistically whether a correlation exists between two variables, x and y, we use the correlation coefficient represented by r. Using Excel it is very easy to plot a scattergraph, determine a correlation between variables and demonstrate the relationship between them by inserting a trendline (where appropriate) between data points. Note that in order for two variables to be correlated, they do not necessarily need to demonstrate a linear trend between them.

Exercise 4.3

The mean radius of lichens growing on gravestones was measured in a churchyard, selecting the largest radius in each case. This was recorded together with the date on the gravestone. The data are presented in Table 4.7. As can be seen from the table, the first task that must be performed

Table 4.7   Mean radius of lichens found on gravestones in a churchyard

Date on gravestone	Mean radius of lichen colony (mm)
1972	2
1963	3
1961	4
1950	20
1937	22
1929	41
1928	35
1920	22
1928	28
1927	35
1917	41
1862	51
1840	35
1918	32

is to place the dates in chronological order. Enter the data into an Excel worksheet and then, using the Sort command from the Data menu, arrange the dates into ascending order (making sure that you select all of the data for sorting). Using Chart Wizard, plot the data and choose the XY Scatter format. Add a suitable title and labels for the *x*- and *y*-axes.

Scattergraphs

In Chart Wizard select the Scattergraph option, XY (Scatter), without lines connecting points. Make sure you edit the scale of axes where points are clustered in one portion of the chart to ensure that all of the points are spread out. This is accomplished by selecting the appropriate axis (*x* or *y*), right clicking the mouse button and from the Format Axis menu selecting the Scale tab. You will then be able to adjust the minimum or maximum value on the axis. To add a trendline to the graph, select one of the points and right click the mouse button. From the options, select Add Trendline. View the different types of trendlines that are available and see how well they fit the points. Options available can be seen in Figure 4.8.

With polynomial and moving average trendlines you may need to adapt the fit of the line by increasing the Order (default value 2).

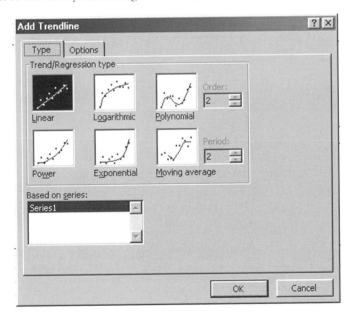

Figure 4.8 Inserting trendlines

Various features of the plot may be formatted. It is usually necessary to edit the thickness of the trendline so that points are not obscured. To format, click on the trendline then change the style and weight of the line to your own preference from the Format Trendline|Patterns menu. From the Format Trendline|Options menu a regression analysis may be performed on the data (see the subsection 'Linear regression') and the line of best fit for the data points inserted into the graph. This is a useful feature where we may want to extrapolate the line. As you can see from Figure 4.9, we can insert the number of units forward or backwards for which the line can be extrapolated on the plot. The equation of the line of best fit to the points may also be inserted by checking the box as shown.

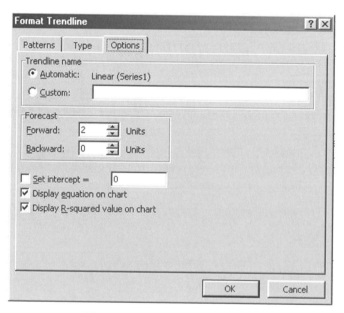

Figure 4.9   Formatting trendlines

Now perform a correlation to see how strong the association is between the two variables: Select Tools/Data Analysis and Click on CORREL from the menu. The CORREL function calculates the product–moment correlation coefficient for the data.

Input the range of cells you want analysed, giving the reference for the dates on the gravestones as the first array and the cell references for lichen size in the second array.

Confirm the selection. A correlation coefficient of −0.75 is obtained. Firstly we should note that the correlation is negative; the more recent the date, the smaller the growth of the lichen colony on the gravestone. The value of 0.75 is midway between 0.5 and 1, so there is a moderately strong relationship between the two variables. As only a small sample has been taken, the data could be supplemented by increasing the number of observations and the correlation recalculated. Using the value of the correlation coefficient alone we are unable to comment on the validity of any relationship between variables. This may, however, be determined from statistical tables, which allows us to decide whether there is a statistically significant correlation between variables at a chosen probability level. The concept of probability in statistical testing is further discussed in section 5.1.

Correlation analysis is frequently performed in medical investigations where we may be looking for the influence of some causative factor upon the incidence of a disease or illness. In many scientific experiments, however, the investigator maintains strict control of a number of variables within an experiment, keeping some variables at a constant level, whereas others may be increased or decreased in order to examine how one variable is dependent upon another (independent) variable. An example might be in an enzyme experiment: the temperature, pressure, pH and enzyme concentration could be kept at constant levels but the concentration of substrate varied to determine the effect upon the rate of catalysis by the enzyme. Where we are interested in examining the relationship of a dependent variable upon an independent variable we must use regression analysis.

Linear regression

Simple linear regression

Where a scattergraph shows points approximate to a straight line, simple linear regression may be used to determine the relationship between two variables. The purpose of the analysis is to place a line of best fit between all of the points and determine how closely the line fits through the points using the 'least squares' method. If all of the points fit the line then the deviations from the line would be 0, but how far they lie away from the line gives an indication as to how well the model fits our observations. The regression coefficient provides us with a regression coefficient, R-squared. If all of the points were to fit the line without any deviation then the R-squared value would be 1; the closer values are to 0, the less likely there is any relationship between variables. Regression uses residual analysis to demonstrate the clustering of observations to the line, where residuals are the observed value minus the predicted values. Examining residuals helps to identify any outlier values that sometimes occur where erroneous values in a data set may be a consequence of sampling or experimental error. We may then decide to omit the outlier from further analysis.

Exercise 4.4

The most frequent use of linear regression analysis in the laboratory is for the determination of a line of best fit through a calibration curve. The R-squared value is used to confirm a linear relationship between x and y and justifies the use of the calculated equation for the line of best fit for the determination of values of x from observed values of y.

During a research project a student was required to make a determination of the protein concentration of an enzyme they were attempting to purify. The student constructed a calibration curve by the Lowry method before attempting to quantify the unknown protein concentration. The results are shown in Table 4.8.

Enter the data onto your Excel worksheet in *two* columns. (This will mean replicating concentration three times, so use Copy and Paste to do this efficiently. You will only need one set of labels, for Concentration and Absorbance, at the top of each column.)

N.B. The experiment was performed in triplicate but it is *not* appropriate to use mean values.

A calibration curve should reflect the variation in the experimental technique, and the analysis should be used to identify any outlier values, so all the replicates must be included.

Table 4.8   Protein determination using the Lowry Assay

| Concentration (µg/ml) | Absorbance | | |
	Replicate 1	Replicate 2	Replicate 3
20	0.106	0.108	0.109
40	0.204	0.202	0.205
60	0.311	0.310	0.311
80	0.417	0.419	0.425
100	0.508	0.510	0.509
120	0.612	0.616	0.614
140	0.722	0.734	0.729
150	0.809	0.819	0.822

To perform the regression analysis select Tools|Data Analysis and highlight Regression from the list. A pop-up box appears in which to enter the range of the data and select some options for the analysis as shown in Figure 4.10.

Input the range of the Y (absorbance) data and then the range of the X (concentration) data. Include data labels in this selection and tick Labels in the Regression box.

Under the Output options, click on the New Worksheet ply to enter the results of the regression analysis on a new worksheet. Select both Line Fit Plots and Residuals then confirm your selections by clicking on OK.

Excel analyses the relationship between independent and dependent variables and produces a report and charts on a new page in your workbook. You may need to move some of the statistics around on the worksheet, together with the charts to be able to see all of the information. The results of the analysis are shown in Figure 4.11. The most important statistic from the analysis is the R square (R^2) value. This indicates how strong a relationship exists between the dependent and independent variables. As the value is 0.997 there is clearly a very strong relationship between concentration and absorbance. The results also show an ANOVA table (see section 5.3 for further explanation of analysis of variance) from which the probability value is used to confirm whether there is a significant relationship between x and y. The P value from the table (shown under the heading Significance F) shows there is a

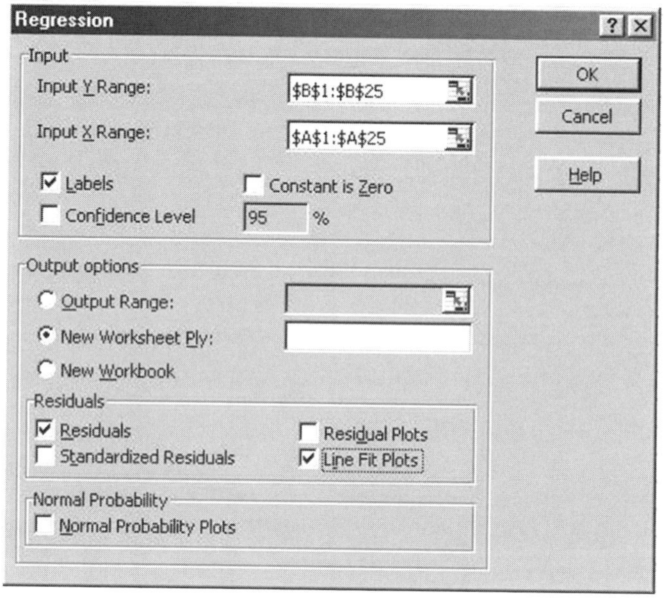

Figure 4.10   Inserting cell ranges for regression analysis

highly significant relationship between absorbance and concentration as $P = 8.19 \times 10^{-29}$, and this value is well below 0.05,
the level of significance adopted. See section 5.1 for a full
explanation of interpreting a level of significance in statistical
tests.

The line plot produced for the data shows individual data
points and (usually in pink) the values of Y (absorbance) that
are calculated as part of the analysis. You will also find these
listed in a table at the bottom of the worksheet. The predicted Y
values on the graph would be more appropriately substituted
by a line of best fit through the observations. Highlight one of
the predicted Y values and right click the mouse button, then
choose Clear to remove them from the chart. Now select one of
the observed values and insert a linear trendline as described
in the previous exercise. The Residual plot shows the clustering
of the observed values around the line of best fit: some are
above and some are below the line; but there are no values
which might be regarded as outliers (some distance from the
baseline). The R^2 value produced in the analysis confirms that

SUMMARY OUTPUT								
Regression Statistics								
Multiple R	0.998354446							
R Square	0.996711601							
Adjusted R Square	0.996562128							
Standard Error	0.01403066							
Observations	24							
ANOVA								
	df	*SS*	*MS*	*F*	*Significance F*			
Regression	1	1.312695051	1.3127	6668.19	8.18994E-29			
Residual	22	0.004330907	0.0002					
Total	23	1.317025958						
	Coefficients	*Standard Error*	*t Stat*	*P-value*	*Lower 95%*	*Upper 95%*	*Lower 95.0%*	*Upper 95.0%*
Intercept	-0.00794431	0.006447876	-1.23208	0.23093	-0.0213164	0.005428	-0.021316	0.00542778
Conc (µg/ml)	0.005315335	6.50919E-05	81.659	8.2E-29	0.005180343	0.00545	0.0051803	0.005450327

Figure 4.11   Regression analysis output table

there is very little scatter about the trendline as this value ($R^2 = 0.997$) is very close to 1.

Another important feature of the analysis is that we are provided with the equation for the line of best fit through the points. A straight line may be described by the equation:

$$y = mx + c \qquad \text{(Equation 4.6)}$$

where m is the slope of the line and c is the intercept through the y-axis. The equation may be used to predict values of x and y (to which confidence limits may be attached) providing R^2 and P values confirm a significant relationship between variables, which in this example they clearly do.

From the table produced on the worksheet the value for the intercept is seen in the Coefficients column; this value is -0.0079 (refer to Figure 4.11). The slope is beneath this, next to Conc (µg/ml); the value is 0.0053. If we then substitute these values in Equation 4.1, we arrive at the equation for the line of best fit through our data points:

$$y = 0.0053x - 0.0079 \qquad \text{(Equation 4.7)}$$

Where the analysis becomes useful is in determining unknown concentrations of protein (x) after measuring an absorbance value (y). Instead of extrapolating the value of y

from a hand-drawn plot where the line of best fit has been drawn in by placing a ruler onto the plot and determining the best place to draw the line manually, we now have a more accurate means of quantitating our unknown value. This means rearranging Equation 4.2 to solve for x:

$$x = (y + 0.0079)/0.0053 \qquad \text{(Equation 4.8)}$$

So if we had an absorbance reading of 0.1 then, if we substitute this in the rearranged equation, we should obtain a value of 20.4 μg/ml for concentration.

Multiple regression

In the previous sections we have investigated the relationship between one (independent) variable and another (dependent variable). There may be times; however, when we suspect that there is a relationship between more than two variables and that these are interdependent. To determine how to relate these variables we must use multiple regression.

In simple linear regression we demonstrated the relationship between x and y as:

$$y = mx + c \qquad \text{(Equation 4.6)}$$

In multiple regression we imply that y is linearly dependent on one variable (x_1) and also dependent on another variable (x_2), so:

$$y = m_1x_1 + m_2x_2 + c \qquad \text{(Equation 4.9)}$$

This equation assumes that the dependent variable, y, is dependent on two independent variables, x_1 and x_2. m_1 and m_2 are partial regression coefficients because they can reflect how a value of y would change with a unit change of x_1 if x_2 were held constant, and vice versa. Where y is dependent on more than one variable, then the equation may be adapted to include as many variables as necessary. So if y is dependent on four variables then:

$$y = m_1x_1 + m_2x_2 + m_3x_3 + m_4x_4 + c \qquad \text{(Equation 4.10)}$$

In multiple regression we are able to obtain an equation from which we are able to predict y from values of x_1, x_2, etc. and so develop an understanding of which variables are able to affect y. This is a useful function for exploring complex relationships as within living systems it is unusual to find that an association is restricted to just two variables.

Exercise 4.5

The systolic blood pressure of an individual is thought to be related to a person's age and weight. Table 4.9 shows the age, weight and systolic blood pressure for a sample of eight healthy subjects. Enter the data as shown onto an Excel worksheet.

Note that the dependent variable (systolic blood pressure), y, is kept on the right in one column; the independent variables (x_1 and x_2, age and weight) are kept together on the left. As in the previous exercise, from the Tools|Data Analysis menu highlight Regression from the drop down menu. In the dialogue box:

1 for Input Y range: type in the cell references for the column that contains the independent values (systolic BP) including the title.

2 for Input X range: type in the cell references for the columns containing all of the dependent variables (the two remaining columns), again including titles.

3 In the dialogue box, click on Labels, Residuals, Residual Plots, and Line Fit Plots.

Table 4.9   Age, weight and systolic blood pressure in eight healthy subjects

Age (years)	Weight (kg)	Systolic BP (mmHg)
50	77.3	130
53	79.5	135
56	81.8	140
59	84.0	145
60	88.6	150
62	90.9	155
65	93.2	160
70	97.7	165

4 In Output options, type in a cell reference on your worksheet where you would like the statistics to appear and confirm your selection with OK.

A complete analysis of the multiple regression model should now appear on your worksheet.

Interpretation of the regression analysis

The R-squared value of 0.992 indicates that there is a relationship between the variables and that systolic blood pressure may be explained using a linear model, where age and weight are explanatory variables. The residual plots are a useful check as to whether the assumption of linear regression is appropriate. The output from Excel gives residual plots for each of the variables. As may be seen from the output, each of the comparisons shows that the points are clustered around the central line. If there was no likelihood of a relationship between variables, then the points would show a purely random scatter.

Using the TREND function

If we are satisfied that the regression analysis demonstrates a relationship and that the resulting equation can be used as a model, then if there were four subjects of known age and weight, it could be useful to predict what their systolic BP would be.

Enter the following values on your worksheet underneath the columns for Age and Weight: (leave a few rows blank between these theoretical values and your actual data).

Age (years)	Weight (kg)
54	71.2
55	71.2
56	71.2
57	71.2

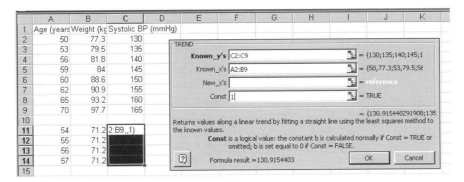

	A	B	C	D	E	F	G	H	I	J	K
1	Age (years	Weight (kg	Systolic BP (mmHg)								
2	50	77.3	130								
3	53	79.5	135								
4	56	81.8	140								
5	59	84	145								
6	60	88.6	150								
7	62	90.9	155								
8	65	93.2	160								
9	70	97.7	165								
10											
11	54	71.2	2:B9,,1)								
12	55	71.2									
13	56	71.2									
14	57	71.2									
15											

TREND
Known_y's C2:C9   = {130;135;140;145;1
Known_x's A2:B9   = {50,77.3;53,79.5;56
New_x's   = reference
Const 1   = TRUE
                                  = {130.915440291908;135
Returns values along a linear trend by fitting a straight line using the least squares method to the known values.
    Const is a logical value: the constant b is calculated normally if Const = TRUE or omitted; b is set equal to 0 if Const = FALSE.

Formula result = 130.9154403   OK   Cancel

Figure 4.12 Using the TREND function in Excel

Choose a group of cells to contain the predicted (SBP) values (the four cells to the right of those just used for the theoretical values would be the most logical) and select them.

Click on the Paste Function button and choose TREND from the Statistical list. The TREND box appears in which you are prompted to enter the raw data and the range of cells containing the information for which you require predictions made (this function can be also be applied in simple linear regression), as shown in Figure 4.12.

Type in the ranges on your sheet that contain your observed y-values (SBP), the observed x-values (age and weight). This time do not include the labels.

In the box labelled 'const' type in 1 (meaning True). (This confirms that an intercept term is required for the equation describing the relationship between the variables.) Then click Finish.

Now move to the rows that were selected for inputting the predicted values.

Press the Function key, F2. The word Edit should appear on your status bar at the bottom of the screen. Hold down both Control and Shift keys and press Enter. The formula bar should now display the TREND function and the cell references for the observed and predicted values, and the predicted values should appear in the selected cells.

The values are based on a best-guess prediction, where a 95 per cent prediction interval uses the best guess plus or minus two standard errors of the estimate. We can therefore be 95 per cent confident that the systolic blood pressure will lie in this range.

WEB SUPPORT – SECTION FOUR

Here you will find some examples to work through to look at the shape of distributions and calculate the appropriate descriptive statistics. There will also be some exercises to work through on correlation and regression. Worked solutions will be available for all of the exercises.

5

Statistical Analysis

So far we have considered how as part of a scientific investigation we design experiments based on previous research in which we test our interpretations that are formulated into a hypothesis. As part of the design process the most appropriate statistical analysis for the data should be considered, keeping our plan for the investigation as simple as possible. In this section we look at the most commonly used statistical tests and how we may apply them using Excel.

5.1 Selecting a statistical test

Before starting a plan of work, we have to consider very carefully the design of the experiment to ensure that we are conducting a fair test. At the end of the experiment we use a statistical test in order to establish whether or not our hypothesis can be accepted. The purpose of applying statistical tests to experimental data is to determine whether there is a significant difference in our observations that is, to examine the probability that our samples are different.

Probability

Probability is a means of quantifying the likelihood of a particular event taking place. By an event we mean the result of an experiment that is of particular

Data Analysis and Presentation Skills by Jackie Willis.
© 2004 John Wiley & Sons, Ltd   ISBN 0470852739 (cased)   ISBN 0470852747 (paperback)

interest. In conducting the experiment we are gathering data in order to determine the outcome of the investigation. In designing our study we have to make sure that we do not introduce any bias into the investigation so that the outcome is measured as fairly as possible. This frequently means ensuring that the sequence in which samples are taken (trials) are performed in a random order. By performing a number of trials we are able to gather information on the probability of an event taking place.

If we were to toss a coin 50 times and record the result of each toss (heads or tales), we could determine the number of heads recorded for each 10 tosses. We would expect that our chances of obtaining heads would be 50:50, that is there is a 1 in 2 probability (0.5 expressed as a decimal) of obtaining heads.

During the course of the experiment we would see that as the number of trials increases, the chance of obtaining heads gets closer and closer to 0.5.

From the experiment we can say that the probability of being able to toss a head is:

$$\frac{\text{number of events}}{\text{number of trials}} = 0.5$$

If the probability of an event occuring is P then the probability of it not happening is $(1 - P)$, i.e. the probability of obtaining tails with tossing the coin is $(1-0.5)$. Probability is frequently converted into a percentage, so the probability of tossing a head is 50 per cent.

Exercise 5.1

Seventy seeds were scattered on agar in a petri dish and kept in the dark at 15°C for 14 days. At the end of this period 37 seedlings were observed. What is the probability of the seeds germinating under these conditions?

i.e. 37/70 = 0.53 (53%)

Calculating probability

We can use the formula bar in Excel to calculate this probability, and convert it into a percentage:

Open a new workbook in Excel.

Click on an empty cell on the Excel spreadsheet.

Enter the formula $= 37/70$.

Press the Enter key and the probability will appear on your worksheet (0.5287).

If we want to modify the formula to show the percentage, then we must click on the cell again and adjust the formula to read $= (37/70) * 100$.

We would conclude that the probability of seeds germinating under the specified conditions is 53 per cent.

The probability that the seeds will not germinate is $1 - 0.5287 = 0.4714$, which is the same as saying $(70 - 37)/70$, so the probability of the seeds not germinating is 47 per cent.

In choosing which type of statistical test is best for our data we need to consider, at the planning stage, the characteristics of data that we are going to collect.

There are a number of statistical tests that can be used to determine whether there is a significant difference between *two* samples. These are the:

- Z-test for independent samples
- Z-test for paired (matched) samples
- *t*-test for independent samples
- *t*-test for paired (matched) samples
- Mann–Whitney U-test
- Wilcoxon signed rank test
- Chi-squared test (see section 5.4).

In order to decide which is the most appropriate we have to take account of a number of factors about the data that we are dealing with.

Types of data

Data can be described as continuous or discrete.

By continuous data we mean that data have been quantified in some way. Its accuracy will be dependent on the precision with which it has been measured. For example, we may have used the Lowry method to determine the amount of protein in a given sample. We may then report its protein content, but the number of decimal places that we would choose to use to report the value is dependent on the precision of the analytical technique.

With discrete data we are dealing with exact numbers, usually determined by a counting method. This could be the number of petals on a flower, heart rate, or cells counted using a haemocytometer. In each case we are dealing with exact numbers, so we would have 6 petals, 60 heartbeats per minute or 12 cells in a grid.

In each of these two examples, data is numerical and has been measured or counted and therefore has definitive values. These data are also known as interval data.

The statistical tests that are applied to interval data are the Z-test and the Student t-test.

Not all data generated in an experiment is precise in this way. Sometimes we may need to consider variables more difficult to quantify, such as an emotional response or the severity of a disease. This type of variable cannot be measured accurately; this type of data is known as ordinal data. Statistical tests that may be applied to ordinal data are the Mann–Whitney U-test or the Wilcoxon signed rank test.

In certain experiments we may need to collect information that is descriptive about the subjects in our investigation. Where data are descriptive, we tend to summarize the information by placing it into different categories. Examples of categorical data include eye or hair colour, species within a genus, or male/female subjects. Data that are categorical are also known as nominal data. The Chi-squared test is applied to data at the nominal level.

Independent and paired samples

In planning an experiment we try to eradicate as many sources of variation as possible by limiting the number of factors likely to influence our results. This sometimes involves generating what are known as matched or paired samples. Where data are paired, the test variable is measured within the same experimental subject or sample. By providing information from the same subject it is possible to eliminate variability that may occur between samples and so each individual will act as their own control. Data that are not matched or paired are independent.

Characteristics of the sample population

The choice of test used will depend upon the characteristics of the population from which the sample is taken, i.e. whether it is normally distributed, skewed or bimodal. In section 4.2 we considered normal distributions and deviations from normality. In some instances we will know the shape of the population (e.g. heights of individuals are normally distributed) or are able to make the assumption that it is normally distributed on the basis of comparison with similar distributions. More usually the shape of the population is unknown but, providing the sample taken is large enough, it may be possible to assume that it is representative of the rest of the population and is normally distributed. It is also possible to test whether data complies with a normal distribution. The Chi-squared goodness of fit test described in section 5.4 may be applied to test for normality.

The size of the sample

The larger a sample, the more representative it will be of the population from which it has been taken. If a slight significant difference exists between the mean values of two populations, a test that includes a large number of samples will be more sensitive to detect this difference than one involving a small number of samples. As already discussed in section 2.2, we have to ensure that the size of sample used in an investigation is large enough to prevent a Type I error occurring, otherwise small differences will remain undetected. At the same time we have to be aware that there may be environmental or resource issues that enter into a decision about sample size.

5.2 Statistical tests for two samples

For samples that contain more than 30 subjects, the Z-test is usually preferred. Biological investigations quite frequently involve small samples. Under these circumstances it is important to know something about the shape of the distribution of the population from which the sample has been taken. Where it appears that the data approximate to a normal distribution (follow a typical bell-shaped curve) then the t-test is generally used. Where the shape of the sample deviates from a normal distribution, i.e. is skewed, or there is uncertainty about the shape of the population, the Mann–Whitney or Wilcoxon signed rank test would be applied.

Table 5.1 Statistical tests for matched or independent samples

Sample size	Distribution	Matched samples	Independent samples
$n > 30$	Normal or skewed	Z-test (matched) paired samples	Z-test independent samples
$n < 30$	Normal	Paired t-test	Independent t-test
$n < 30$	Normal or skewed	Wilcoxon signed rank test	Mann–Whitney U-test

Standard deviation of the population

In most instances the standard deviation of the population can only be determined from the sample data. If the samples are large, the estimates of the standard deviation should be reliable and the Z-test may be used (irrespective of the shape of the population).

If the samples are small (less than 30), estimates will be poor and the t-test should be used, providing the samples indicate a normal distribution.

Table 5.1 provides a summary of the factors that need to be considered when choosing a statistical test.

The Wilcoxon and Mann–Whitney tests are known as non-parametric tests because, unlike the t-tests, they may be used on data that may or may not follow a normal distribution (distribution free). The t-tests are therefore known as parametric tests as they may only be applied where the data is known to comply with normality.

Hypothesis testing

Hypotheses are used by investigators to define the purpose of their experiment. For a hypothesis to be accepted it must be tested; on the basis of the test results the hypothesis is either supported or rejected, or may need to be modified.

The null hypothesis and the direction of the alternative hypothesis

In statistical analysis we formulate a null hypothesis (H_0) for our experiment and it stands against the alternative hypothesis (H_1). The null hypothesis makes an assumption that the factor under investigation has no effect; whereas the alternative hypothesis is formulated on the assumption that the factor does have an effect.

In considering how we should state the alternative hypothesis we have to reflect on the scientific evidence on which our experiment is based, as we need to determine whether there is only one possible outcome in our investigation or more than one outcome. The tests that may be applied are either one-tailed or two-tailed.

If the alternative hypothesis specifies a direction (i.e. there can only be one significant consequence of our experiment), then a one-tailed test is used. If the alternative hypothesis does not have a direction (i.e. more than one outcome is possible) then a two-tailed test is used.

Example of a one-tailed test

A chemical additive in a cosmetic is considered to have carcinogenic properties. In an experiment to determine whether this can be confirmed a group of rats have the cosmetic applied to their skin. A control group of rats (numbers equal in each group) have the same cosmetic applied but without the suspect chemical present. Each group of rats is monitored for the appearance of malignant growths.

In this experiment there is only one possible outcome for the experiment, either the rats will develop tumours or they will not, so there is only one possible direction that is being tested. We would therefore adopt the one-tailed test for our alternative hypothesis.

Example of a two-tailed test

If we were investigating the effects of a new chemical being developed as a fertilizer for tomato plants, in the absence of any previous work, we would be unsure of what effect the substance might have on the growth of the plants. If we wanted to design an experiment to determine the effects of the chemical we would probably start by taking a group of plants and dividing them into two sets. One set would form a control group that would not be exposed to the chemical, whilst the other would be grown under identical conditions but with the chemical applied. After a set period of time we would examine the two samples to establish whether the growth of the plants had been altered.

It may be that the chemical is effective and promotes plant growth or that it proves ineffective or maybe even brings about the stunting of growth. In this experiment we cannot be certain of the direction of the outcome and so it is appropriate to adopt the non-directional two-tailed test. The null hypothesis would therefore state that there would not be any difference in plant growth between control and treated sets of plants; the alternative hypothesis would propose that there is a difference in plant growth but would not specify whether growth would be likely to increase or decrease.

A cautious approach should be taken when considering whether to adopt a one-tailed test; the majority of statistical analyses are performed using a two-tailed test. Once an experiment has been completed, the direction of a change in the test variable compared with a control is sometimes very clear from looking at the results. It should be decided before the experiment has taken place that it is appropriate for a one-tailed test to be adopted; this must be on the basis of scientific evidence that there can only be one direction, i.e. one possible outcome for the experiment, if the results prove to be significant.

Level of significance

In a statistical analysis we are testing our certainty of accepting the null hypothesis. Before a test is performed, the level of significance for the rejection of the null hypothesis must be decided. Although the level of significance can be set to any value, it is usually set at 5 per cent ($P < 0.05$). This means that the likelihood of the event taking place by chance alone is 5 or less in 100 (so there is at least a 95 per cent probability that the null hypothesis is correct), i.e. it is very unlikely to take place by chance alone. The lower the level of significance that is adopted, the less likely it is that the null hypothesis will be rejected.

Presentation of a statistical test

Using Excel for statistical analysis makes it easy to write on the worksheet the full basis of the test being adopted and the conclusions that may be drawn from the analysis. The hypotheses and details of the tests applied to data should always be clearly stated, as should a description of the results and your conclusions from the analysis. You may find it useful to use the following checklist for each analysis you perform.

1. State the null and alternative hypotheses, indicating a direction to the alternative hypothesis if appropriate.
2. Indicate whether the test is one-tailed or two-tailed.

3. Provide the name of the test applied (and assumptions about the population from which the samples to be tested are drawn).

4. Set the level of significance at which the null hypothesis will be rejected (normally $P < 0.05$).

5. Input the data into a table on the worksheet and apply the test.

6. State the outcome of the statistical analysis, i.e. whether the null or alternative hypothesis is accepted, together with the level of significance found in the test.

7. Comment on the data, i.e. what the test has shown (e.g. an increase in plant growth using the fertilizer with an mean increase of 15 per cent in the size of tomatoes). It may also be pertinent to comment on the quality of the data used or variability found in the experiment.

Using the statistical functions in Excel

Statistical tests may be accessed through the Data Analysis functions from the Tools menu. The computer that you are working on may not already have these functions available, so before you commence your analyses:

Click on Tools: Data Analysis

If the Data Analysis functions do not appear at the bottom of the drop-down menu then:

Click on Tools: Add-Ins, then click on the Analysis ToolPak from the checklist that appears. The Data Analysis option should now appear when you click on Tools. (N.B. You may need the original CD that Microsoft Excel was loaded from if you are accessing the software from your computer's hard drive.)

Writing on your worksheet

When typing information and comments on your worksheet, it is easier to use textboxes than to type directly into cells. Using the textbox will prevent some of the text becoming 'hidden' within the cells and prevents text overflowing from one page to another, which you can sometimes be unaware of unless the document has been formatted carefully before printing.

To use textboxes click on the Draw icon on the toolbar and then click on the Textbox icon to enter your comments.

The Student *t*-test for independent samples

Exercise 5.2

An investigation was conducted on the effects of dietary fat in margarine on serum cholesterol concentrations in male subjects. In a controlled experiment 12 male subjects were given a diet that used a standard 'low fat' margarine. A separate group of 12 male subjects were provided with a diet that substituted a new type of margarine reported to significantly lower serum cholesterol in comparison with other brands. The subjects were given the diet to follow for six months, then their serum cholesterol levels were compared. In the first group one subject discontinued the study, leaving only 11 subjects. The serum cholesterol values for the two groups of subjects are compared in Table 5.2.

The independent *t*-test was adopted for the data of Table 5.2 on the following basis:

1. The serum cholesterol of the subjects is measured at the interval level.

2. The subjects from both groups were chosen at random from a population of medical students, and so were not matched with one another (ruling out the use of a paired test).

3. Previous experiments have demonstrated that the serum concentrations of cholesterol in humans is normally distributed and this assumption was made about the test subjects.

Table 5.2   Serum cholesterol concentrations in subjects after 6 months on different dietary regimens

	Serum cholesterol concentration (mg/dl)										
'Low fat' margarine	175	168	154	163	171	134	149	151	147	155	162
'New' margarine	139	145	165	132	170	144	136	162	159	161	168	168

4. The range of the distributions for the two groups were not widely different, so the standard deviations of the groups are unlikely to be dissimilar.

'Low fat' margarine: Range $= 175 - 134 = 41$ mg/dl
'New' margarine: Range $= 170 - 132 = 38$ mg/dl

(The standard deviations needed estimating).

5. The size of each sample is less than 30, making it appropriate to use a t-test for independent samples.

Null hypothesis: The type of margarine provided in the diet of the test subjects has no effect on their serum cholesterol concentrations.

Alternative hypothesis: The type of margarine provided in the diet of the test subjects does have an effect on their serum cholesterol concentrations.

Test: Two-tailed t-test for independent samples. (We cannot be sure of the direction of the outcome as the new margarine may cause serum concentrations to increase or decrease, or remain unaffected.)

Level of significance: $P = 0.05$ (5 per cent).

We can now perform the test using the Data Analysis option in Excel.

Enter the data onto the worksheet in two columns as shown below:

Low fat margarine	New margarine
175	139
168	145
154	165
163	132
171	170
134	144
149	136
151	162
147	159
155	161
162	168
	168

Select the Data Analysis function from the Tools menu.

Choose *t*-test: Two Sample Assuming Equal Variances from the menu.

A dialogue box as shown in Figure 5.1 will appear in which to input the cell references for each column of data as the Variable 1 range and Variable 2 range. Include the rows that have the titles for your data in your selection and tick the check box that shows Labels. Ensure Alpha is set at 0.05, the default value, which is the level of probability that will be adopted for the test. (0.05 means that you are assuming the 5 per cent level of probability.)

Now select where you would like your results to appear in the workbook. If you do not alter this to your current worksheet then the analysis will appear on a separate sheet. It is usually more convenient to select an empty cell below your data table. To accomplish this, click on Output Range and either select a cell on your worksheet or type in the cell reference (e.g. B15).

Click on OK and the statistical analysis will be summarized on the worksheet as shown in Figure 5.2. We now need to examine results of the analysis and comment on the outcome of the test.

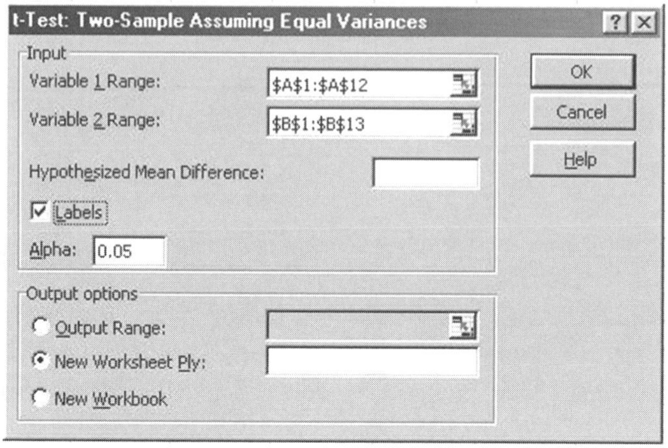

Figure 5.1   Entering data for the independent *t*-test

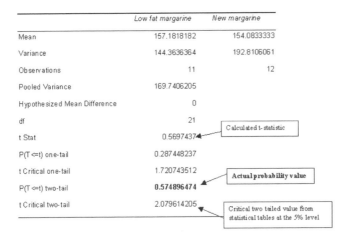

	Low fat margarine	New margarine
Mean	157.1818182	154.0833333
Variance	144.3636364	192.8106061
Observations	11	12
Pooled Variance	169.7406205	
Hypothesized Mean Difference	0	
df	21	
t Stat	0.5697437	Calculated t-statistic
P(T<=t) one-tail	0.287448237	
t Critical one-tail	1.720743512	
P(T<=t) two-tail	**0.574896474**	Actual probability value
t Critical two-tail	2.079614205	Critical two tailed value from statistical tables at the 5% level

Figure 5.2   Output data for the independent *t*-test

Interpretation of the statistical analysis

If we had performed the statistical analysis manually, we would have followed a set formula that would give us a calculated *t*-statistic (labelled as t-Stat in Excel). As we can see from the table, this value is 0.569 743 7. We then need to refer to a set of tables for the Student *t*-distribution to find what is known as the critical value that determines whether or not our data are statistically significant at the 5 per cent level. In order to look up the appropriate value, we need to know the degrees of freedom (df) for the data. The degrees of freedom for the Student *t*-test for independent samples is $n-2$, so, as there are 23 observations in this example, df$=23-2=21$. In the Appendix you will find the table for the Student *t*-distribution. Find the two-tailed critical value for 21 degrees of freedom. The value should be 2.0796. As you will see by comparing the results table in Excel, this value is already provided, as is the critical value for the one-tailed test (1.7207).

In order to accept the alternative hypothesis, the calculated *t*-statistic should be greater than the critical value. Clearly in our example this is not the case as 0.569 74 $<$ 2.0796. If you look at the results table again, then you can see that above the critical two-tail value Excel has returned the actual probability value for the analysis. This value is 0.5748, i.e. the test has proved there is no significant difference between the two margarine diets that were used, as the level of significance from the analysis is 57.5 per cent. We therefore accept the null hypothesis that there is no difference in the cholesterol levels for the subjects taking the two dietary treatments.

Conclusion: A comparison of the mean data for the two groups indicates that the mean serum cholesterol concentration for the subjects following the 'low fat' margarine diet was 157 mg/dl (with a variance of 144 mg/dl) in comparison to the subjects who followed the diet with the new margarine whose mean serum cholesterol concentration was 154 mg/dl (with a higher variance of 193 mg/dl). As the P value for the analysis was 0.575, we can conclude that the null hypothesis may be rejected and the alternative hypothesis accepted. From the experiment we have shown that the type of margarine used had no effect on serum cholesterol concentrations for the participants of the study.

Although a significant result has not been obtained, the experiment has not PROVED conclusively that the dietary margarine did not have an effect on serum cholesterol. The experiment was performed once, in a small number of subjects. Although this gives weight to the argument that the diet did not have any effect, the more times the experiment is repeated and a similar result obtained, the *more likely* it is that the hypothesis is correct. We could also improve upon the design of the experiment by using each subject as their own control. We have no indication of the serum cholesterol concentrations at the start of the experiment, before the diet was begun, to know whether either diet caused a change in cholesterol levels in each subject.

Student *t*-test for dependent (matched/paired) samples

Maintaining variability at as low a level as possible is an important consideration in the design of experiments. One means of minimizing variability is to design an experiment on a paired or matched basis. Imagine we want to examine the efficacy of a new 'long-acting' formulation of aspirin (Z) with a standard compressed tablet preparation (Y). We could recruit eight patients who would be willing to participate in the experiment, but there is likely to be many factors that vary within the patient group – they are all not going to be the same height, weight or age, have the same state of health or have symptoms of exactly the same severity. What rules can we apply to the experimental conditions to ensure that these factors are minimized?

1. Each patient can have administered, on separate occasions, the new formulation and the standard aspirin preparation. As the assessment of the efficacy of the treatment will be carried out by the patients themselves, any intra-subject variability will be eliminated by generating matched data.

2. Bias may be removed from the experiment by adopting a double-blind technique. The order in which the preparations are administered can be randomized (four patients will receive aspirin on the first occasion, whilst the remaining four will receive the new drug) and the experiment will be double-blind. A double-blind design means that both treatments will be coded (Y or Z) so that neither the patient receiving the medication nor the doctor giving the tablets will be able to identify which treatment is being given. The code for the treatment is kept by a third, independent party. Section 2.2 discusses study designs to eliminate bias.

At the end of the experiment the investigator will have an assessment of the number of hours of pain relief from the patients. In the experiment we have generated paired data as the subjects have acted as their own control. The paired *t*-test can therefore be used to analyse the data.

Exercise 5.3

The results of the experiment can be seen in Table 5.3.
   Open a new workbook in Excel and enter the data, as in the last exercise, in two columns. The assumptions about the test,

Table 5.3 Pain relief in eight patients administered standard aspirin tablets and a new drug on two separate occasions as part of a double-blind study

Patient	Hours of pain relief with standard formulation (Y)	Hours of relief with new formulation (Z)
1	3.2	3.8
2	1.6	1.8
3	5.7	8.4
4	2.8	3.6
5	5.5	5.9
6	1.2	3.5
7	6.1	7.3
8	2.9	4.8

reason for using a paired analysis and hypotheses should be included on the worksheet. We will be adopting a two-tailed test as before as we cannot be certain as to whether the new formulation will increase or bring about a decrease in the hours of pain relief in the patients.

When this has been completed, from the Data Analysis menu select *t*-Test: Paired Two Sample for Means. A dialogue box should appear similar to that in Figure 5.1. Input the range of cells for the data for each column under Variable 1 range and Variable 2 range. Include the rows that have the titles for your data and tick the check box Labels.

Ensure that Alpha is set at 0.05 and choose where on the worksheet you would like the results of the analysis to appear. Click OK to confirm your choices.

The data analysis table in Figure 5.3 should now be shown on the worksheet.

From the analysis table we can see that there are a few differences from the previous test results. Firstly, if we were calculating the *t*-statistic using the set formula we would need to subtract individual values in each column from each other as the analysis uses the differences between pairs. This has resulted in a negative value being returned for the calculated *t*-statistic. We ignore the negative sign, as it is only the numerical

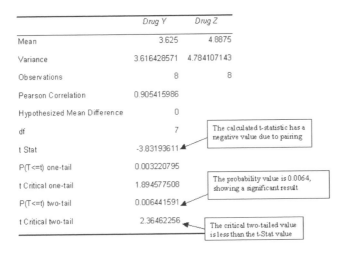

	Drug Y	Drug Z
Mean	3.625	4.8875
Variance	3.616428571	4.784107143
Observations	8	8
Pearson Correlation	0.905415986	
Hypothesized Mean Difference	0	
df	7	
t Stat	-3.83193611	
P(T<=t) one-tail	0.003220795	
t Critical one-tail	1.894577508	
P(T<=t) two-tail	0.006441591	
t Critical two-tail	2.36462256	

The calculated t-statistic has a negative value due to pairing

The probability value is 0.0064, showing a significant result

The critical two-tailed value is less than the t-Stat value

Figure 5.3   Output data for the dependent (paired) t-test

value that we use (if we had our data organized with the column of values for Z first on the worksheet, then Y, we would have a positive value for t-Stat, but the numerical value will still remain as 3.8319).

The calculation of the degrees of freedom is also different. For the paired t-test the degrees of freedom is equal to the number of pairs of data minus one, i.e. $df = 8 - 1 = 7$.

Comparing the calculated value of the t-statistic with the critical two-tailed value at the 5 per cent level of significance, we can see that the calculated value is higher than the tabulated value $(3.832 > 2.364)$. We can conclude that there is a significant difference in the hours of pain relief produced by the new formulation Z compared with the standard aspirin preparation Y and therefore reject the null hypothesis and accept the alternative. As before, Excel shows the actual significance level which is 0.0064 (0.64 per cent). We may make a full statement about the conclusions of the analysis by comparing the means and variance of the data as in the first exercise.

Non-parametric tests for two samples

These tests are used where we have either ordinal data or interval level data from populations which are not normally distributed (or their shape is unknown). When using summary statistics to describe the results from non-parametric tests is it more appropriate to use median values rather than the mean (that is used for parametric tests).

- The Wilcoxon signed rank test is used for matched or paired samples.

- The Mann–Whitney U-test is used for independent samples.

Neither of these tests can be performed automatically in Excel through the Data Analysis options, but making use of the functions on the worksheet the appropriate statistics can easily be obtained.

The Wilcoxon signed rank test

The sign test uses information on the direction of differences between data in pairs and, by ranking the data, the magnitude of the differences is also taken into consideration. We will look at an example where patients suffering from rheumatoid arthritis were asked to grade their joint stiffness after one month of treatment having taken a standard treatment compared with a new drug, in a randomized double-blind study. The patients were asked to record the degree of stiffness in the affected joints immediately upon waking in the morning and rate it on a scale between 0 and 5, where 0 indicates no stiffness and 5 represents complete immobility. As the patient's scores are likely to be subjective it is important that paired data are obtained and that a non-parametric test is applied. The Wilcoxon signed rank test is chosen as this is for matched data.

Exercise 5.4

Enter the data from Table 5.4 in two columns on the Excel worksheet as in previous tests. State the basis of the test:

Null hypothesis: There is no difference in the scores for joint stiffness in the patients taking the standard treatment compared with the new drug.

Table 5.4 Scores recorded for joint stiffness in a group of 10 patients

Patient number	Standard treatment	New drug
1	3	3
2	4	1
3	4	2
4	2	2
5	0	1
6	1	3
7	2	2
8	1	2
9	3	2
10	3	1

Alternative hypothesis: There is a difference in the scores for joint stiffness in the patients taking the standard treatment compared with the new drug.

Level of significance: 5 per cent.

The test is two-tailed as we cannot be certain that the new compound will improve joint mobility.

In order to complete the Wilcoxon test you will need to work through steps 1–6 below.

Step 1

As the data are paired, the first step is to take the difference between each pair. Label a new column next to 'New drug' called 'Difference'.

In the first cell enter a formula to calculate the difference between the scores for each treatment for patient 1, i.e. if your first row of data begins in B2, then type in $=B2-C2$ and press Enter. An answer of 0 should now appear in cell D2. Using the Autofill handle (see section 3.1) copy the formula down the column to calculate the differences between the remaining pairs of data. Your worksheet should now appear as in Figure 5.4.

Figure 5.4   Data table for the Wilcoxon signed rank test

Figure 5.5   Adding signs to the Wilcoxon signed rank test

Step 2

Type the title 'Sign' in the column next to 'Difference'. You will now record the Sign (+ or −) of the differences. Where a sign is negative a value of −1 will be entered, where a sign is positive 1 is entered. Click on first cell in the Sign column (cell E2 in Figure 5.4). From the Paste Function select SIGN and enter the cell reference (D2). A 0 will appear as there is no sign attached to a value of 0, but if you use the Autofill handle to copy the sign function down the column, values will appear in the other cells. Compare your worksheet with Figure 5.5.

Step 3

We now need to use the difference between each pair, but remove any negative values, as in the next stage of the analysis the differences will be ranked. The simplest way to accomplish this is to multiply the Difference by the Sign to return positive values for all the differences. Enter the title 'Sign × difference' in the next column and in the cell below enter a formula to multiply the value in the first cell in the Difference column (D2) by the Sign (0), i.e. in this example $=D2*E2$. The value of 0 should be returned which can then be copied down into the remaining cells.

Step 4

The next stage is to sort the data so that they may be ranked. When data from a table is sorted, ALL of the data in the table has to be selected. If this approach is not taken, then the sorting process will scramble the data.

Select the data, i.e. all rows and columns containing data on the worksheet including labels. Using the Data|Sort command select Sign × difference from the drop down menu as in Figure 5.6

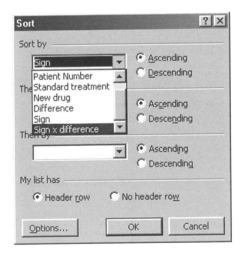

Figure 5.6   Sorting data for Sign × difference

A	B	C	D	E	F	G
Patient Number	Standard treatment	New drug	Difference	Sign	Sign x difference	Rank
1	3	3	0	0	0	
4	2	2	0	0	0	
7	2	2	0	0	0	
5	0	1	-1	-1	1	
8	1	2	-1	-1	1	
9	3	2	1	1	1	
3	4	2	2	1	2	
6	1	3	-2	-1	2	
10	3	1	2	1	2	
2	4	1	3	1	3	

Figure 5.7   Preparing to rank data for the Wilcoxon signed rank test

and sort the data into ascending order. The worksheet should have the data listed as in Figure 5.7.

Add the title Rank to the column next to Sign × difference (G2). The data need to be ranked manually as there are some rules to be applied when ranking data. Firstly, there are three values of 0. It is a rule for the test that any zero differences between the pairs are excluded from the analysis. The ranking should therefore start with the first (and lowest) value of 1. However, there are three Sign × difference values of 1. We therefore have to consider these values as occupying ranking positions 1, 2 and 3 (N.B. If all of the values were different then these would be the ranks assigned.) Because the values are identical we have to award 'tied ranks' to give them equal weighting in the analysis. A tied rank is the average value of ranks, so the average of ranks 1, 2 and 3 will be 2. Next to the three values of 1 enter ranks of 2. We are now ready to continue ranking. The following values in Sign × difference are three values of 2. These will occupy ranking positions 4, 5 and 6; as the mean of these is 5 we enter this value in the Rank column. The last value is 3, so this occupies rank 7.

Ranking

Both the Wilcoxon and the Mann–Whitney tests use ranked data.

If two or more values are the same in the list to be ranked, give each value the mean of the ranks they occupy (as in the example).

Any differences of 0 should not be ranked. You should ignore any zero differences.

Step 5

Now the Signed Rank needs to be calculated. (This indicates the direction of your data, so brings back the + or − status of the differences.) Enter the title Signed Rank into cell H1. The sign of the rank is calculated by multiplying the Sign by the Rank value (by applying the formula = G5*E5 in the example shown). Using the Autofill handle, copy the formula down the column.

Step 6

In the final step we separate positive ranks from negative ranks from which we will calculate the totals of each set (the lower of these two totals will be used as the critical value of T, the Wilcoxon statistic).

Using Data|Sort, sort all of the Signed Rank values into ascending order. This will group all of the positive and negative values together. Separate positive ranks from negative ranks and calculate the totals using the AutoSum function. (To do this you can use the copy button. Select the first cell where you want the data to appear and then Edit: Paste Special, choosing Paste Values from the list.) Your worksheet should now have the totals for each column as shown in Figure 5.8.

If we compare values for the sums of the positive and negative ranks, the negative ranks total is smaller (9). Whichever value is the smaller (regardless of its sign) is taken to be the calculated value (T). Now refer to the table of critical values for the Wilcoxon signed rank test in the

Patient Number	Standard treatment	New drug	Difference	Sign	Sign x difference	Rank	Signed Rank	Negative ranks	Positive Ranks
6	1	3	-2	-1	2	5	-5	-5	2
5	0	1	-1	-1	1	2	-2	-2	5
8	1	2	-1	-1	1	2	-2	-2	5
9	3	2	1	1	1	2	2		7
3	4	2	2	1	2	5	5		
10	3	1	2	1	2	5	5		
2	4	1	3	1	3	7	7		
1	3	3	0	0	0				
4	2	2	0	0	0				
7	2	2	0	0	0				
							Totals	-9	19

Figure 5.8   Separating positive and negative ranks

Appendix. If the calculated value for *T* is smaller than the critical value then we would reject the null hypothesis. From the table we can see that the critical value for seven pairs of data is 2 at the 5 per cent level (note that although there are 10 subjects in the study we exclude any pairs where the difference was zero). Our calculated value is greater than the critical value, therefore, we reject the alternative hypothesis and accept the null hypothesis for the experiment. We can conclude that there is no apparent difference perceived by the patients in relieving the symptoms of morning stiffness by the new drug.

The Mann–Whitney U test

The Mann–Whitney test is the non-parametric test used for independent data, and may be conducted with unequal or equal sample sizes. In the example given here, sample sizes are unequal, but procedures are exactly the same for equal sample sizes.

Exercise 5.5

A team of investigators wanted to investigate the claim that a particular technique could be used to improve memory. They took two groups of subjects of similar ages and educational

Table 5.5   Number of words recalled by two groups of subjects, one of which was given training in the application of a memory technique

Control group	Treated group
26	15
14	45
32	44
25	41
19	25
15	37
31	42
33	26
29	36
26	14
37	27
29	44
23	41
30	26
	28
	33

ability and subjected each group to a test in which they were given a list of 50 items on a list to memorize. One group was provided with a 1-hour session before the test in which they were given training in the technique. The data from the experiment are listed in Table 5.5.

Null hypothesis: Training in a memory technique does not have any effect on the ability of subjects to recall a list of 50 items.

Alternative hypothesis: Training in a memory technique does have the effect of improving the ability of subjects to recall a list of items.

Level of significance: 2.5 per cent.

A one-tailed test is used as the researchers have predicted the direction of the outcome (i.e. the memory of the test subjects could not have been impaired by the training technique).

We will now apply the Mann–Whitney non-parametric test for independent variables to the test data.

Step 1

Open a new worksheet in Excel. Enter the data from Table 5.5 onto your worksheet, but place it in two columns as shown in Figure 5.9 so that in the first column a code is applied to indicate whether the subject's data belong in the control (c) group or the trained (t) group.

Step 2

The data now needs to be ranked applying the same principles as the Wilcoxon test; but first we must sort the data. Highlight the cells containing the data and then select Data|Sort. Sort Items Recalled in Ascending order. Enter Rank into the cell next to Items Recalled. Now give a numerical rank to all of the data, keeping in mind that if values are identical, the mean rank should be entered.

Step 3

The two data sets are separated into control and treated groups once more. To do this we perform another sort, this time selecting to sort the Group alphabetically (select all cells containing data on the worksheet and sort using the Alphabetical Sort button on the toolbar). The two data sets now need to be separated.

Select the data for the treated subjects ($n = 16$) and copy and move the treated data as a block into the three columns adjacent to the control values as shown in Figure 5.10. (Highlight the three columns to be moved and drag on the border to achieve this.)

Using the AutoSum button calculate the sum of ranks for both control and treated groups.

Step 4

Using the sum of ranks we can calculate the value of U (the Mann–Whitney statistic) from the formula:

Group	Items recalled
c	14
c	15
c	19
c	23
c	25
c	26
c	26
c	29
c	29
c	30
c	31
c	32
c	33
c	37
t	14
t	15
t	25
t	26
t	26
t	27
t	28
t	33
t	36
t	37
t	41
t	41
t	42
t	44
t	44
t	45

Figure 5.9   Data for the Mann–Whitney test

$$U = n_1 n_2 + n_1 \frac{(n_1 + 1)}{2} - R \qquad \text{(Equation 5.1)}$$

where n_1 is the number in the smallest sample
n_2 is the number in the largest sample
R is the sum of ranks of the *smaller* data set
This formula can be placed into a cell on the worksheet.

$$= (14 * 16) + (14 * (14 + 1)/2) - 173.5$$

Group	Items recalled	Rank	Group	Items recalled	Rank
c	14	1.5	t	14	1.5
c	15	3.5	t	15	3.5
c	19	5	t	25	7.5
c	23	6	t	26	10.5
c	25	7.5	t	26	10.5
c	26	10.5	t	27	13
c	26	10.5	t	28	14
c	29	15.5	t	33	20.5
c	29	15.5	t	36	22
c	30	17	t	37	23.5
c	31	18	t	41	25.5
c	32	19	t	41	25.5
c	33	20.5	t	42	27
c	37	23.5	t	44	28.5
		173.5	t	44	28.5
			t	45	30
					291.5

Figure 5.10   Totalling the rank data for the control and treated groups

On pressing the Enter key a value of 155.5 for U should be returned.

Step 5

In order to complete the test we need to calculate the value for U' using the formula:

$$U' = n_1 n_2 - U \qquad \text{(Equation 5.2)}$$

Enter the formula, $(14 * 16) - 155.5$, into an empty cell. The value of 68.5 should be returned for U'.

Of the two values, U and U', we take whichever is the smaller. In this example U' is smaller with a value of 68.5.

We now need to look up the critical value of U using the Mann–Whitney U tables in the Appendix, using n_1 and n_2. The null hypothesis is rejected and the alternative accepted where the smaller value of U or U' (whichever applies) is *less than or equal to* the critical value of U. As the critical value U is 64 we can reject the alternative hypothesis and accept the null hypothesis (the calculated value of U' is greater than the critical value). The memory training clearly did not have any effect on the number of items recalled by the subjects (median for the control group is 29 items; median for the trained subjects is 34.5 items).

N.B. Both non-parametric tests, the Wilcoxon signed-rank test and Mann–Whitney U-test, rely on the calculated value being smaller than the critical value to demonstrate significance.

5.3 Analysis of variance

In the previous section we have considered how to test data when we have two samples. Quite frequently though, we design an experiment in which we make multiple comparisons as there are several treatments or conditions applied that we want to compare with a control. Including more than one treatment minimizes any variation that might be encountered by conducting several smaller experiments or investigations over a longer period of time, such as seasonal variations, differences in batches of reagents etc., not to mention the cost and resource implications.

As part of designing an experiment where a number of different treatments are applied, an appropriate statistical test needs to be considered at the planning stage. This is particularly important where the experiment is quite complex to ensure that a 'balanced' design is achieved. In balancing an experiment we ensure that there will be either equal replication into groups (known as blocks) or treatments, or that there will be equal precision in the comparison of variables that are investigated. When an experiment is properly balanced we can expect to apply the simplest statistical analysis from which to demonstrate our conclusions with clarity and unambiguity.

Let us think through an experiment in which we want to investigate the effects of several different concentrations of a growth hormone on the growth of plant sections. To conduct a fair test we need to include a control in which the media used for containing the plants would not have any growth hormone present. Then, instead of designing a series of experiments in which a single concentration of hormone would be investigated against a control, we would design an experiment in which several different concentrations were compared simultaneously. So we may have a design in which we have:

	Treatment			
Control	A	B	C	D

in which treatments A–D would each represent a different concentration of hormone. Having completed the experiment you might then be tempted to use a Student *t*-test to compare the growth of plant sections grown in the control media with each individual treatment in turn. This would be incorrect, unless special conditions were applied, and in statistical terms is known as making a Type I error. This is because we would need to make several analyses to be able to compare all of the data to know whether a difference exists between the control and different concentrations, and also whether the concentration itself contributes to the extent that the plant section grows. In order to make a full set of comparisons we would need to perform *t*-tests as follows:

Control vs A	A vs B	B vs C	C vs D
Control vs B	A vs C	B vs D	
Control vs C	A vs D		
Control vs D			

This would mean performing 10 tests in total. It may be difficult to understand why it would be wrong to make this number of comparisons. The answer is that we might falsely conclude that there is a significant difference between treatments as a consequence of performing so many statistical tests, and not on account of any real difference in the data. When we form the basis of a statistical test we formulate hypotheses and set a level of significance, usually at 5 per cent. This means that we are accepting the probability of an event occurring by chance alone is 5 per cent or less. To try to visualize the problem, let us think of a the probability in a statistical test as being a box containing 100 balls. Five of the balls will be red; the remaining 95 white. If we remove a ball from the box (in conducting the test) then there is a chance of 5 per cent that the ball will be red. If the ball that is removed is white and we then remove another ball (testing the data again) then there is an increased likelihood of obtaining a red ball (5 out of 99). This situation is analogous to performing two tests on the same data set. So in our situation, where 10 comparisons need to be made, the possibility of demonstrating a false significant difference would increase each time the test is performed. A Type I error is therefore a situation in which we wrongly assume that there is a difference between samples and incorrectly reject the null hypothesis.

  There are two ways in which we could obviate this happening. The first would be to lower the level of significance, minimizing the opportunity of a false positive result; but then we may commit a Type II error as a real significant difference may be missed. The second option is to use an analysis of variance (ANOVA) test which removes any possibility of a Type I error as the data is tested all together; this option is preferred as the ANOVA is also a more powerful statistical test in this situation. The power of a test is defined by its ability to correctly reject the null hypothesis under the conditions set.

The analysis of variance uses all of the data and returns a single probability value. Should this show that there is a significant difference between treatments in the data set then further analysis, in the form of a multiple range test has to be performed. According to the design of the investigation, either a one-way (for one-factor comparisons) or a two-way (for two-factor comparisons) ANOVA test can be applied. In the following subsections we will look at examples of each of these tests and demonstrate how a multiple range test can be applied.

One-way analysis of variance

This analysis is applied for one-factor comparisons; so for the comparison of the growth of the plant sections, the only factor investigated was hormone concentration. In this situation a one-way analysis would be suitable. However, we will take as our example to work through in Excel an experiment in which the effect of pH on drug dissolution was investigated. A preparation of aspirin containing 100 mg of drug was placed in solutions of different pH for a period of 12 hours in a rotating basket. Samples from each solution were taken at periodic intervals and at the end of the experiment the amount of drug that had dissolved was calculated. The experiment was repeated five times at each pH. The purpose of the experiment was to examine whether the pH of the dissolution medium had any effect on drug dissolution and if so, to indicate at which pH optimum dissolution occurred.

Exercise 5.6

Enter the data in Figure 5.11 on the Excel worksheet as shown.
  Before commencing the test, the null and alternative hypotheses need to be stated, together with the level of significance to be adopted:

Null hypothesis: There is no difference in the dissolution of the aspirin tablets in different pH solutions.

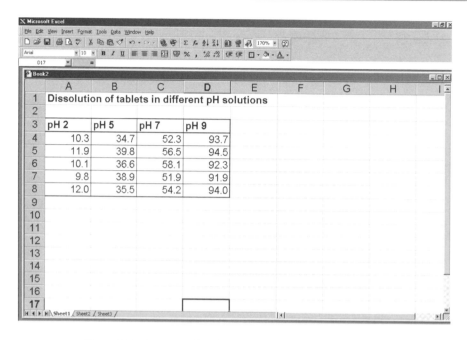

Figure 5.11   Inputting data for the one-way analysis of variance

Alternative hypothesis: There is a difference in the dissolution of aspirin when exposed to different pH solutions.

Level of significance: 5 per cent ($P<0.05$).

Using the Tools|Data Analysis function, select the ANOVA: Single Factor (this is the one-way analysis of variance) option from the menu. As shown in Figure 5.12, enter the range of values for the cells containing the data and check the box for the labels so that the pH will be displayed in the results of the analysis. Our data is in columns, so ensure that this is also checked on the dialogue box before clicking OK.

The analysis for the data will now appear on the worksheet, as shown in Figure 5.13. Excel produces the mean and variance for each set of results at different pH values. From the ANOVA table the value of F can be seen to be 1676.2, greatly in excess of the Critical value of F at the 5 per cent level of significance (3.2). As in the other statistical tests that we have used so far,

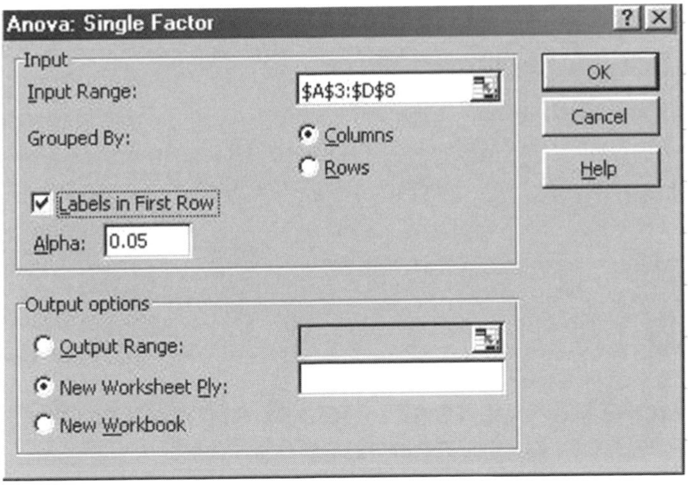

Figure 5.12   Showing the data range for the one-way ANOVA

statistical tables do not need to be used as the *P* value is automatically calculated; on this occasion it is 3.41×10^{-20} (shown as 3.41E-20), i.e. *P* = 0.000 000 000 000 000 000 03. This demonstrates that there is a highly significant difference between the dissolution of the drug at different pH, so we can therefore reject the null hypothesis and accept the alternative hypothesis. The problem now arises in deciding how the pH treatments differ; there are six comparisons to make in total and we have already indicated that we cannot do a pairwise comparison using the *t*-test. How do we test for differences without increasing the likelihood of a Type I error? The answer is to use a multiple range test. There are a number of different types: Tukey, Scheffé and least significant difference (LSD) between means test. They all work in the same way; it is only the situations in which they are applied that differ slightly. We will apply a LSD between means test to the data to determine what differences lie in dissolution rates at each pH. This test may only be applied where there is an equal number of samples in each treatment.

Least significant difference (LSD) analysis

Using this test we are able to compare all of the differences between mean values in our data set and determine what the lowest value for the difference between any pair of means would need to be for there to be significance at a given level. The steps in the calculation of the LSD may be seen below.

Multiple range test – least significant difference between means test

1. The first step is to calculate the standard error of the difference between any two group means from the formula:

   s.e. $= \sqrt{\{\text{mean square within groups } [(1/n) + (1/n)]\}}$ (Equation 5.3)

   where the mean square (MS) within groups has been calculated in the analysis of variance and is shown in the ANOVA table under Sources of Variation Within Groups, and n is the number of observations in each group.

   So for our example:

   from the ANOVA table the mean square within samples (groups) $= 3.571$ and $n = 5$
   therefore
   $$\text{s.e.} = \sqrt{\{3.571[(1/5)+(1/5)]\}}$$

   so this would be calculated in Excel from the formula:

   $$= \text{SQRT}(3.571*(1/5+1/5))$$

   Having entered the formula into an active cell the value of 1.195 should be returned.

2. We now use the s.e. to find what the least difference between means will be for various levels of significance. From the ANOVA table the degrees of freedom (df) associated with the mean square within groups is 19 (calculated on the basis that there were five observations in each group and four treatments, so df $= (5 \times 4) - 1)$.

Using the table of critical values for the Student *t*-test in the Appendix, look up the 5 per cent and 1 per cent points of the *t*-distribution for 19 df. You should find that these are 2.093 and 2.861 respectively.

The LSD is calculated by multiplying the s.e. by each value, therefore the smallest difference between means at the:

5 per cent level will be 2.5 (2.093 × 1.195) and at the

1 per cent level will be 3.4 (2.861 × 1.195).

In order to find out where significant differences are we must take each set of means for each pH and subtract differences. Using the facilities of the Excel spreadsheet it is easier to rank mean values and then make pairwise contrasts as shown in Figure 5.13. Using the LSD data we can determine where significant differences exist between each pair of means. (In order to report this fully, you may want to calculate the least significant difference at a range of probability levels, 5, 1, 0.5, 0.1 per cent, as appropriate.)

We can now make some comparisons. For there to be a difference in drug dissolution at the 5 per cent level of significance there needs to be a minimum difference between

ANOVA						
Source of Variation	SS	df	MS	F	P-value	F crit
Between Groups	17957	3	5985.65	1676.18	3.4E-20	3.23887
Within Groups	57.136	16	3.571			
Total	18014.1	19				
	s.e.=	1.19532				

Least Significant Difference between Means Analysis

pH	2	5	7	9	
	10.8	37.1	54.6	93.2	mean
2	-	-	-	-	
5	26.3	-	-	-	
7	43.8	17.5	-	-	
9	82.4	56.1	38.6	-	

Figure 5.13   One-way ANOVA and least significant difference between means analysis

means of 2.5 and at the 1 per cent a difference of 3.4. From these comparisons we can clearly see that there is a significant difference in means which can be summarized as follows:

The drug dissolution at pH 2 is less than that at pH 5, 7 or 9.

The drug dissolution at pH 5 is less than that at pH 7 and 9 but more that at pH 2.

The drug dissolution at pH 7 is less than that at pH 9 but more than that at pH 2 and 5.

The conditions for drug dissolution are optimum at pH 9 as dissolution is greater than at pH 2, 5 or 7.

(N.B. Unless there is found to be a significant difference in treatments shown in the ANOVA, there is no justification in then continuing and performing the LSD test.)

Two-way analysis of variance with replication

In the two-way ANOVA with replication we examine the effects of two treatments (factors) with replication in each treatment. For example, in the above experiment we may have conducted our tests with two different formulations of the drug, in which case we would be looking at both the effect of the drug formulation and the effects of pH on drug dissolution. We will work through an exercise in which we will make comparisons of two factors using the two-way ANOVA.

Exercise 5.7

In a Phase I clinical trial the pharmacokinetics of a new drug was investigated in young and elderly subjects. An oral dose of the drug was given as a single dose and blood specimens were collected for 12 hours; dosage was then continued twice daily for a period of two weeks after which the trial subjects attended and blood samples were taken as before. The area under the drug concentration time curve (AUC) was calculated for each subject for Days 1 and 15 of the trial. The data need to be examined to determine whether:

- there was any significant difference in AUC for Day 1 and Day 15
- there was any significant difference in AUC between young and elderly subjects

Before starting the statistical analysis we need to state the hypotheses for the investigation. We are examining two factors so we need to consider both of these when formulating the hypotheses.

Null hypothesis: *This will be a statement that there will not be any significant difference in either of the two factors investigated.*

There is no difference in the AUC between Days 1 and 15 of the study, or between young and elderly subjects.

Alternative hypothesis: *There are two alternatives that can be considered here, either one or both may be found to be true if the test demonstrates a significant difference.*

There is a significant difference in the AUC for the drug comparing a single dose at Day 1 with a period of multiple dosing on Day 15.

There is a difference in the AUC between young and elderly subjects.

Enter the data in Figure 5.14 onto your worksheet, including the labels as shown. The two-way ANOVA is accessed through the Tools|Data Analysis menu. From the list provided highlight Anova: Two-Factor With Replication. Enter the cell references containing the data in the Input Range box, making sure that you also include the labels. In the Rows per Sample box type 8 as there are data for eight subjects, both young and elderly, on each study day. Set the level of significance, α, to 0.05, then click OK.

Two Way Analysis of Variance		
	Day 1	Day 15
Young	1006	1602
	1205	1756
	1385	1312
	1726	1652
	850	1482
	866	1026
	1425	1672
	1272	1761
Elderly	1944	1522
	1730	1984
	1807	2172
	1689	1582
	1686	1378
	1429	1148
	2594	2079
	2225	2188

Figure 5.14   Inputting data for the two-way ANOVA with replication

The worksheet should now contain the ANOVA table that will show the Average values (and their associated variances) for the young and elderly subjects on Days 1 and 15 of the study, and the AUCs for young and elderly subjects combined. The ANOVA table may be seen in Figure 5.15. This time, as distinct from the one-way analysis, there are three probability values.

The first, defined as Sample, is a value of 0.00075 and represents the between-rows analysis, i.e. the probability that AUCs for young and elderly subjects are different. As the probability is below 0.05 we can confirm that there is a significant difference between AUCs and by comparing mean values state that AUCs in the elderly subjects are higher, so it would appear that elderly subjects handle the drug differently from younger subjects.

The second probability value in the Columns row represents the between-columns analysis for young and elderly subjects combined, so that any difference between AUCs on Day 1 and Day 15 may be determined. The value of 0.44 shows that there is no significant difference between the two days, so the drug would not appear to accumulate after two weeks' dosing using this regimen.

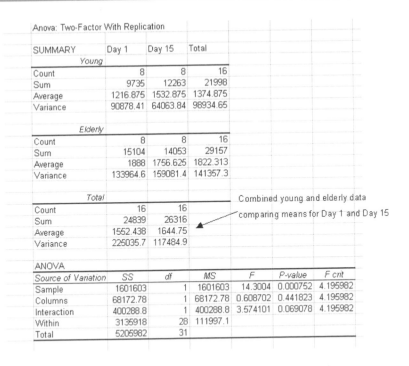

Figure 5.15   Summary output for the two-way ANOVA with replication

The final probability level is labelled Interaction and takes into account both factors (age and multiple dosing). The probability for Interaction can be used to determine whether there is an interaction between the two variables, age and multiple dosing, or if the effect of each variable is additive. The P value of 0.07 would indicate that there is no significant difference in AUC caused by the age of the subjects during multiple dosing. If a significant interaction were found, this might suggest a significant accumulation of the drug due to the advanced age of the subjects and limit the use of the drug owing to safety issues. As the value is close to 0.05 it might be questionable as to whether the sample size was sufficiently large to be certain that there was no effect. A fair amount of variability is also evident in the data.

Two-way analysis of variance without replication

This test is also known as the ANOVA using a randomized block design and like the previous test examines two factors within an experiment. A block is a set of data that has been grouped by the experimenter to allow very little variation within the block, before being randomized to particular treatments. There may be some variation between blocks due to various external factors, but, as the data within the block is more consistent, grouping the data in this way will help to minimize experimental error. As previously discussed, the experimental plan should ensure that a balanced design has been devised so that blocks are comparable for the analysis. When an experiment is balanced we can expect to apply the simplest statistical analysis from which to state our conclusions with clarity and without ambiguity.

Exercise 5.8

In an experiment to determine whether pretreating seeds by refrigeration causes an increase in germination, seeds were assigned to two treatments: control, where seeds were kept under normal environmental conditions for 4 weeks before planting, and cold-treated where seeds were kept for four weeks at 4°C. Seeds were sown in batches of 50 (equivalent to blocks) over a period of 12 months. The growth of the plants after 6 weeks was compared and the mean growth for each batch calculated.

For each batch sown the environmental conditions will be consistent; each batch represents a block. Between batches there may have been some local variation in conditions, in which case we must test the data not only for the difference in treatments but for differences between blocks. The data may be analysed using the two-way ANOVA without replication that will determine whether there is a difference in the germination of the plants and if this is influenced by external factors.

The data is entered onto the worksheet as shown in Figure 5.16 Select Tools|Data Analysis and from the dialogue box highlight Anova: Two-Factor Without Replication and click OK.

Plant germination experiment - mean plant heights (cm)		
Batch	Control	Refrigerated
1	50.6	52.2
2	51.3	50.4
3	60	57.4
4	61.1	62.4
5	62.7	61.9
6	59.9	58.7
7	58.6	56.2
8	51.3	50.9
9	52.7	53.6
10	53.4	54.1
11	49.8	50
12	47.5	49.5

Figure 5.16   Data for the two-way ANOVA without replication

In the Input Range box type in the cell references for your data (including the labels and column giving the batch numbers). Check the Labels box to indicate that you have done this. Click on OK. The ANOVA table should now appear on your worksheet as shown in Figure 5.17. There are two probability values, one showing the probability of a difference between rows, the other the probablity of a difference between columns (but unlike the two-way analysis with replication there is no interaction between rows and columns).

The analysis for the growth data demonstrates the following:

- differences between batches/blocks (rows $P= 0.000\,000\,26$), therefore there is a difference in the rate of germination of the plants in the different time periods that the seeds were sown, most likely due to seasonal changes affecting growth.

- no difference between treatments (columns, $P=0.76$), therefore there is no difference in the growth of the plants depending on the prior treatment of the seeds before sowing.

Anova: Two-Factor Without Replication				
SUMMARY	Count	Sum	Average	Variance
1	2	102.8	51.4	1.28
2	2	101.7	50.85	0.405
3	2	117.4	58.7	3.38
4	2	123.5	61.75	0.845
5	2	124.6	62.3	0.32
6	2	118.6	59.3	0.72
7	2	114.8	57.4	2.88
8	2	102.2	51.1	0.08
9	2	106.3	53.15	0.405
10	2	107.5	53.75	0.245
11	2	99.8	49.9	0.02
12	2	97	48.5	2
Control	12	658.9	54.90833	26.93174
Refrigerated	12	657.3	54.775	20.58932

ANOVA						
Source of Variation	SS	df	MS	F	P-value	F crit
Rows	510.2583	11	46.38712	40.90794	2.63E-07	2.817927
Columns	0.106667	1	0.106667	0.094067	0.764796	4.844338
Error	12.47333	11	1.133939			
Total	522.8383	23				

Figure 5.17   Summary output for the two-way ANOVA without replication

5.4 The Chi-squared (χ^2) test

In the previous sections we have looked at data where we were examining differences between means or medians. In this section we will explore the use of the Chi-squared test that is used when data from one or more samples has been placed into categories, i.e. the data are nominal. Data can vary in complexity according to the observations taken in an investigation and so the way in which it is applied is adapted for each situation.

Basis of the test

In the Chi-squared test we usually want to know if there is a difference between observations that have been recorded and sorted into different categories. As with any other statistical test we formulate a null and an alternative hypothesis. In the Chi-squared test we are interested in finding whether the frequency of our observations is in line with what we expected (reflected in a statement of the null hypothesis, that there will not be any difference in observed and expected frequencies), or whether a different pattern has emerged during the investigation (reflected in the statement for the alternative hypothesis that there will be a difference in observed and expected frequencies). The test is two-tailed as we do not specify in which direction we would expect any change in frequencies to occur.

There are a few conditions to the use of the Chi-squared test:

1. Only frequency data can be compared using the test, not percentages or proportions as these do not take into account the size of the sample. Sample size has a direct bearing on the outcome of a test, as in any other type of statistical analysis. Once the test has been performed we can then make comparisons on the relative frequency of events by conversion to percentages or proportions.

2. The test may only be applied where expected frequencies are greater than 5 otherwise any resulting probability value would be invalid.

In the following exercises we will look at three different situations in which the Chi-squared test is used.

Comparing categories in a single sample

This is the simplest situation in which we collect frequency data; observations are made with one sample from which two or more options may be selected. The frequency data shown in Table 5.6 was obtained in an experiment in which the preferences of a sample of students was observed for two different types of chocolate. The frequencies reported are the *observed frequencies* and the data are organized into three categories. The purpose of the experiment was to investigate whether there was a preference by test subjects for milk or dark chocolate or whether their selection was completely random.

Null hypothesis: There is no difference in the number of pieces of milk or dark chocolate selected by the group of students.

Alternative hypothesis: There is a difference in the number of pieces of milk or dark chocolate selected by the group of students.

Level of Significance: 5 per cent ($P < 0.05$).

N.B. The Chi-squared test is always a two-tailed test, so this need not be quoted when performing the test.

Exercise 5.9

Enter the observed frequencies onto your Excel worksheet from Table 5.6.

Use the AutoSum button to calculate the total number of pieces of chocolate consumed.

Although the observed frequencies (number of pieces consumed) is recorded in the experiment, we now need to calculate the expected results, i.e. what results would we expect if the selection of the chocolate was a completely random process? If the process were random, we would expect that it would be equally likely that the number of pieces of chocolate consumed would be exactly the same (like tossing a coin and choosing heads or tails), therefore the probability should be 50:50.

The expected number of pieces eaten will equal

Total number of pieces/2

(as there are two types of chocolate).

On the Excel worksheet calculate the expected consumption using the above relationship, i.e. enter the formula $= (205+289)/2$. An answer of 247 should be returned. If the selection of the chocolate pieces was completely random we would expect that exactly 247 pieces of both dark and milk chocolate would be eaten. We now have to test this against the observed results to find out whether our observations are significantly different from what we expected. Create a second column in the table and enter the expected results as shown in Table 5.7. We are now ready to perform the test.

Click on a cell in the worksheet where you want the result of the test to be reported. The value that is returned is the

Table 5.6   Preference for milk or dark chocolate shown by a test group of subjects

Type of chocolate	Number of pieces consumed by test subjects
Dark chocolate	205
Milk chocolate	289

Table 5.7   Selection of dark or milk chocolate by a group of students

Type of chocolate	Number of pieces consumed by test subjects	Expected consumption of chocolate pieces
Dark chocolate	205	247
Milk chocolate	289	247

probability value only. From the Paste Function menu select CHITEST from the Statistical options. Enter the cell references for the Actual (observed) range, and then for the expected range and confirm your choices. The Chi-squared value of $P = 0.000\,157$ is added to your worksheet. This is less than the set significance level of 5 per cent. We therefore reject the null hypothesis and accept the alternative hypothesis: the selection of the chocolate is not a random process, the test subjects show a preference for milk chocolate.

Goodness of fit test – data from a genetics experiment

Genetics experiments are primarily concerned with predicting the phenotype of various crosses with different genotypes. The Chi-squared test is invaluable for determining whether the outcome of a breeding experiment is in keeping with predicted Mendelian ratios. Mendel gained his reputation for cross-breeding experiments with peas. Some crosses involved the inheritance of more than one characteristic. We will now work through an example where in a cross-breeding experiment with pea plants, plants with round yellow peas were crossed with plants producing wrinkled green peas. The first generation (F1) plants from the cross all had round yellow seeds, indicating that the offspring were all heterozygous, having alleles for both sets of characteristics, but with round and yellow alleles being dominant. The F1 generation were then self-fertilized and in the resulting offspring the following characteristics were observed:

Type of pea	Observed frequency
Round yellow	68
Round green	28
Wrinkled yellow	23
Wrinkled green	10

The predicted Mendelian ratios for the experiment were:

Type of pea	Expected ratio
Round yellow	9
Round green	3
Wrinkled yellow	3
Wrinkled green	1

In these circumstances we will be applying the Chi-squared test to test the *goodness of fit* of the expected results to those observed in the experiment (for this reason the Chi-squared test is sometimes referred to as the *goodness of fit test*).

Exercise 5.10

Enter the observed data and expected ratio onto your worksheet as shown in Figure 5.18. Using the theoretical ratios, we need to work out the expected frequencies for the different types of peas in the experiment. Firstly we need to calculate the total number of pea plants produced from the cross (use the AutoSum feature for this calculation). This should give an answer of 129. The total now needs to be split in the proportions of 9:3:3:1 so the first calculation that needs to be

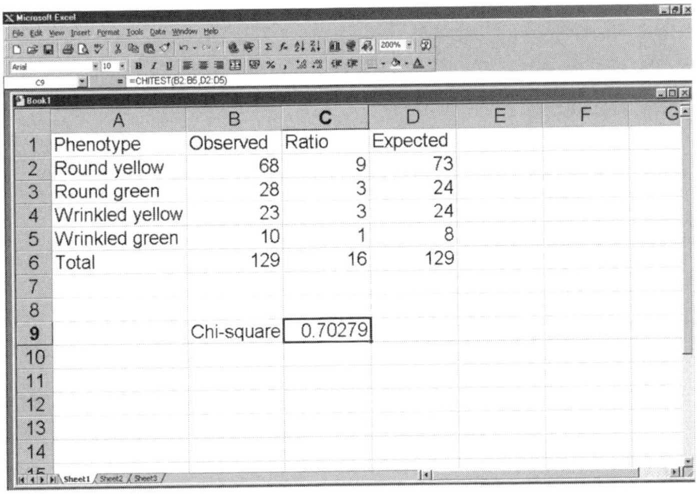

	A	B	C	D
1	Phenotype	Observed	Ratio	Expected
2	Round yellow	68	9	73
3	Round green	28	3	24
4	Wrinkled yellow	23	3	24
5	Wrinkled green	10	1	8
6	Total	129	16	129
9		Chi-square	0.70279	

Figure 5.18 Data for the Chi-squared test comparing Mendelian ratios

made is to find out how many fractions of the total should belong in each category. If we were having to share out a piece of cake in these ratios we would know that we would have to count up how many pieces of cake in total would be required, so the answer to this is to add up the proportions, i.e. 9+3+3+1=16.

The next step in to calculate what 1/16th of the total will represent: i.e. 16 parts=129 peas, so 1 part=129/16=8.0625. (Calculate the answer using Excel.)

Once we have this value the observed frequency can be calculated as:

Expected number of yellow smooth peas = 9 parts (therefore 9 × 8.0625 peas)

The calculation is repeated for the remaining ratios until a complete set of expected frequencies for all of the phenotypes is produced on the worksheet. The Chi-squared test can now be performed. Using the Paste Function, select CHITEST as before. The probability value is entered onto your worksheet and should be 0.703. The results of the test shows that we can accept the null hypothesis as there is no difference in the observed and expected frequencies of the pea plants. We can therefore conclude that the observed frequencies were not significantly different from those predicted by the Mendelian model (we can therefore accept its validity for predicting the genotype of the peas).

Comparing two samples

This version of the Chi-squared test is used when we are comparing the outcome of an experiment involving two samples. It is applied to determine whether a particular treatment has any affect on the outcome of the experiment, where one of our samples is often a control. This test has many applications, particularly where we are interested in finding out whether a treatment or event has had an effect on a population from which samples are taken for comparison.

Exercise 5.11

It is suspected that a component of a weedkiller spray has a deleterious effect on the growth of a particular type of crop. An experiment was set up in which 300 seeds were selected at random and sown under identical conditions in three separate plots with 100 seeds in each. One plot was sprayed with the weedkiller, the second plot was sprayed with a special mixture of the weedkiller in which the suspect component was not added, and the third plot was sprayed with water. The seeds were then raised under identical conditions and the number of seeds that germinated after a period of 1 month were counted in each plot. The results of the experiment can be seen in Table 5.8.

   In order to apply the Chi-squared test we must do as we have in previous examples and calculate the expected frequencies associated with the experiment.

   Enter the data on your Excel worksheet as shown in Figure 5.19, including the blank table in which we will calculate the expected frequencies.

Table 5.8 Comparison of germination in water, special mixture weedkiller and full weedkiller treated plots

	Water treated plot	Special mixture treated plot	Weedkiller treated plot
Germinated successfully	87	91	89
Failed to germinate	13	9	11

Microsoft Excel - Book1

File Edit View Insert Format Tools Data Window Help

K35   =

	A	B	C	D	E	F
1	Observed frequencies					
2						
3		Water treated plot	Special mixture plot	Weedkiller treated plot	TOTAL	
4	Germinated successfully	87	91	89	267	
5	Failed to germinate	13	9	11	33	
6	TOTAL	100	100	100	300	
7						
8	Expected frequencies					
9						
10		Water treated plot	Special mixture plot	Weedkiller treated plot	TOTAL	
11	Germinated successfully					
12	Failed to germinate					
13	TOTAL					
14						
15						

Figure 5.19 Data tables for the Chi-squared test

Firstly, determine the total number of seeds that germinated and did not germinate using the AutoSum button. Using the totals we can calculate the proportion (fraction) of seeds which successfully germinated:

267 out of a possible 300 seeds germinated, so the proportion will be 267/300 (use Excel's formula bar to calculate the proportion). The answer should be 0.89 and the proportion of seeds not germinating will be:

33 out of a possible 300 so this will be 33/300. The answer should be 0.11.

The number of seeds expected to germinate/not germinate now needs to be calculated for treated and control samples (where 100 is the column total): for example,

the number of water treated seeds expected to germinate $= 0.89 \times 100 = 89$

the number of weedkiller treated seeds expected to germinate $= 0.89 \times 100 = 89$

the number of special mixture treated seeds expected to germinate $= 0.89 \times 100 = 89$

(Note that the numbers are identical here as an equal number of seeds was allocated to each treatment in the experiment; this is not always the case.)

We now repeat the calculation for the seeds that are not expected to germinate for each treatment:

the number of water treated seeds not expected to germinate $= 0.11 \times 100 = 11$

the number of weedkiller treated seeds not expected to germinate $= 0.11 \times 100 = 11$

the number of special mixture treated seeds not expected to germinate $= 0.11 \times 100 = 11$

These values may now be inserted into the table for the Expected frequencies. When this has been completed the Chi-squared test can be performed.

Select a cell in which you want the probability value entered and select CHITEST as in the previous examples. Enter the cell references for your observed and expected tables (but do not include row and column totals).

The probability value for this experiment is 0.66. Clearly there is no difference in the numbers of seeds germinating following treatment with the weedkiller with or without the suspect ingredient and so we may conclude that it has no effect upon germination.

Yates' correction (continuity correction)

This is normally applied where there are 2 rows × 2 columns as the number of degrees of freedom (df) for the test becomes 1, or where values in cells of the table are less than 5. The degrees of freedom are calculated from the number of categories that are placed into columns or rows:

$$df = (\text{number of rows} - 1) \times (\text{number of columns} - 1) \qquad (\text{Equation 5.4})$$

where the $df = 1$ Yates' correction is applied to provide a more conservative value for the Chi-squared statistic and to prevent a Type I error from occurring (refer back to section 5.3). There is no function in Excel to automatically calculate Yates' correction and produce a probability value, as in the previous exercises. An example of its use is provided on the book's support website as calculations for Chi-squared with the correction must be made and then compared with a critical value from statistical tables.

WEB SUPPORT – SECTION 5

Plenty of examples will be supplied so that you will be able to practise using the tests discussed in this section. Each will show how the solution has been worked out using the Excel functions and will provide an analysis and interpretation of the results.

6

Presentational Skills

So far we have looked at how we initiate a scientific investigation, plan an experimental design and then present and analyse the results. When these processes are complete we need to communicate what we have found to other scientists. Dissemination usually takes the form of a scientific publication, but as journals can sometimes take several months to produce accepted papers, research is frequently presented at conferences in the form of an oral presentation or poster. The standard software that is used on these occasions is Microsoft PowerPoint, so we will spend some time in exploring how this package works, but firstly let us think about how a presentation should be planned and carried out.

6.1 Preparing for seminars

The most common form of oral presentation is a seminar. These are sessions in which a speaker makes a brief presentation, followed by questions from the audience. In preparing for a seminar you need to plan carefully as strict time limits usually apply and as much information as possible should be conveyed in

Data Analysis and Presentation Skills by Jackie Willis.
© 2004 John Wiley & Sons, Ltd   ISBN 0470852739 (cased)   ISBN 0470852747 (paperback)

a way that will interest and engage your audience. Here are some tips on how to prepare for oral presentations.

Preparing information for the seminar

Research the background information thoroughly and make sure that you understand it before trying to plan your presentation. Look for diagrams, figures and tables that you will be able to adapt and display in your talk (making sure that there is no issue of copyright by using them). Make a list of the bibliography and references as you prepare the written material as any references used should be acknowledged.

The next step is to make some notes for the presentation. Once you have completed this, read through them and see how long it takes. You should allow extra time for explanation of any figures or tables that are included.

Now summarize the notes so that you can prepare the slides/overheads that you plan to use. Keep slides as simple as possible, conveying the message of the main points that you are trying to make in clear scientific language. Avoid lengthy sentences that will clutter the slide; instead use bullet points that are short and easy to read. Choose an appropriate font size, keeping in mind the size of the room and screen where you will be making the presentation.

When preparing slides in PowerPoint, choose the colours of the text and background carefully. You need to select a scheme that has impact, but which can be seen easily at a distance: If you are preparing overheads against a clear background then keep to a black or dark blue for the colour of the text; avoid some of the other popular colours, such as green, red and yellow, as writing becomes very difficult to read if these are used.

Everyone enjoys preparing slides in PowerPoint because of the animated features and sound effects. These should be used with some caution in presentations as they can become a major distraction and eventually an irritation to your audience. Anyone who has dyslexia will also find your presentation very difficult to follow, particularly where blocks of text are animated on a slide.

Make sure that figures, diagrams and tables are properly labelled and titled. If these are taken from a book, paper or website then they should be properly referenced with the source clearly shown on the slide. Any material taken from the Internet should be checked to ensure that there are no copyright restrictions that apply to its use. Avoid scanning in items as these frequently produce very pixelated images. There are many sources of free good quality images on the World Wide Web and by using the Draw features in PowerPoint it is possible to annotate and customize them for your talk.

Your seminar should be clearly structured, so begin with a title page that shows your name and the institution where you are studying. To follow there should be an introduction that provides background information to the topic that you are going to present so that the audience is eased into the more complex information that they may not be familiar with. The main part of your talk should come after the introduction. If you are presenting some research then it is a good idea to present the aims and objectives of your investigation then lead into the design and methodology of your study before moving on to the results and discussion. Every presentation should have a conclusion to summarize the main points that you have presented. References used can be given on a slide at the end or as part of a handout to the audience.

Presenting the seminar

Once the slides or overheads are complete you should have a practice run-through of your seminar. Do this in front of an audience of friends or family; or, if you can't face this, then try a mirror. If you have access to a video camera then ask someone to record your practice seminar. It is important to know how you are presenting yourself while you are talking and to have feedback on your performance. Be aware of your body language; don't slouch in front of the audience, never looking at them. Stand upright, smile and introduce yourself before you start. Avoid folding your arms, instead use your hands to engage with the screen and the audience. When you are explaining something shown on the screen, don't turn your back to the audience. Use a pointer, or your hand if you don't have a pointer, to indicate features of interest, making sure that you stand to the side of the screen and not in front of it. Appear calm and relaxed, even if inside you feel far from it. Taking a few deep breaths before you go up to take your position usually helps to calm the nerves that everyone inevitably feels, particularly when it is a new experience – the first time is always the worst.

When you begin speaking make sure that you attempt to project your voice forward to the audience, maintaining a correct posture will ensure this. During the presentation remind yourself to sustain the level of your voice so that it does not start loud and then trail away towards the end. Make sure that the pace of the presentation is even, without being too slow or rushed. This is where practising beforehand is important to make sure that the timing is correct. Try to vary the way in which you speak, emphasizing important words and phrases, so that you avoid talking in a flat monotone. Avoid um's and ah's when you are speaking as this can become very distracting. If you lose your place then simply start again, but avoid becoming flustered as this will only result in panic. Prompt cards can be useful, but is usually better to use the slides that you have prepared as the prompts so that you do not end up reading the entire seminar from the notes that you have prepared. Once you have finished, make this clear to your audience. This is normally done by thanking the audience for their attention and asking if anyone has any questions. Although you will have researched your area thoroughly it is not expected that you have suddenly become a world authority. If you don't know the answer to a question it is all right to say so, but at the same time try to use the knowledge that you have gained to speculate and form an opinion.

6.2 Using Microsoft PowerPoint

This software package is used for making high-quality presentations. Presentations may be produced using text, clipart, drawing and graphs. The presentation itself is a single file that contains:

- slides
- speaker's notes
- handouts for delegates

Each individual page of the presentation is a slide on which you can display a variety of information and graphics. The slides may then be printed out as black and white or colour transparencies, or made into 35-mm slides. The presentation may also be made using multimedia facilities available in most lecture theatres.

Starting a new presentation in PowerPoint

Enter the program by selecting PowerPoint from the Microsoft Office suite of programs. You will see options from which to choose to open a new presentation via the AutoContent Wizard (uses a set of pre-labelled and formatted slides), a template (blank slides in a preset format), to open a new set of blank slides or to open an existing presentation. If you want to explore presentations that have already been set up for a variety of purposes then browse through the list offered by the AutoContent Wizard and choose one to have a look at. You can add information yourself in order to customize it to your own needs. Save one of the presentations with the filename 'test' and then close the file.

Opening an existing presentation

From File|Open, select the PowerPoint file test.ppt that you have just created (note that the extension .ppt has been added to the file name to indicate that it is a PowerPoint file). Double click on test.ppt to open the presentation.

There are different ways of being able to view your presentation selected using five buttons located near the bottom left of the screen. By moving the mouse, take the pointer over each one in turn to see the ToolTip description of each function. The buttons will be:

Slide View   Outline View   Slider Sorter View   Notes Page View   Slide Show

N.B. Slide View may be called Normal View on some versions of PowerPoint.

Move to the button for Outline View and select this option (by clicking on it). By clicking on the slide number on the left-hand side of the screen you are able to edit information on the slide, whilst on the right of the screen you can see the changes taking effect on the slide. Close the presentation. As a demonstration of how to use PowerPoint for seminar presentations, we are going to produce a short set of slides from which the handout in Figure 6.1 has been prepared.

Starting a new presentation using a template

Now start a new presentation (Click on File|New) and, using the Presentation Designs menu as seen in Figure 6.2, sort through the different presentation templates until you find one that you prefer.

The title slide

Once you have selected your template, click on OK. A series of different types of slides will be shown for you to choose the most appropriate sort for the slide in the presentation sequence. Flick through the different slide types to see what is available (the types are explained in a box on the bottom right of the dialogue box), but then select the Title Slide. Type in the heading 'Vitamin C in *Citrus* Fruits' as shown in slide one in Figure 6.1. We now need to insert clipart and find a file containing an image of a lemon. From the Insert menu select Clip Art from the list provided (noting that you could insert the image direct from a file if you already had something available). From the resident clipart gallery provided you can scroll through the images until you find something

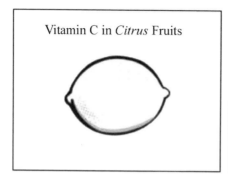

Vitamin C in *Citrus* Fruits

Characteristics

- Edible fruits belonging to the *Citrus* family
- Fruits have a thick rind and juicy pulp
- Grown in regions between 40° north south latitude
- Rich in vitamin C and citric acid, the latter providing their distinctive taste

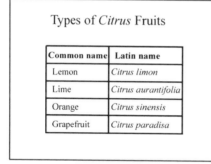

Types of *Citrus* Fruits

Common name	Latin name
Lemon	*Citrus limon*
Lime	*Citrus aurantifolia*
Orange	*Citrus sinensis*
Grapefruit	*Citrus paradisa*

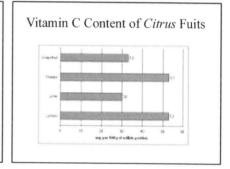

Vitamin C Content of *Citrus* Fuits

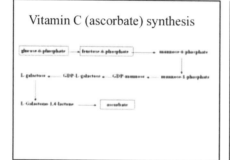

Vitamin C (ascorbate) synthesis

Vitamin C in the Diet

- Humans cannot synthesise vitamin C so it must come from the food we eat
- 60 mg of vitamin C is required daily
- Deficiency causes bleeding gums and loosening of teeth (scurvy), increased susceptibility to infection, weakness and lethargy

Figure 6.1   Sample handout prepared using PowerPoint

suitable and then adding it to the presentation using the Insert button. Images may also be imported from on-line resources. Choosing the clips on-line option will take you through to the Microsoft clipart site on the Internet, but there are many other free on-line resources. An excellent resource for scientific and medical images is provided by the Wellcome Trust library which contains 160 000 images (http://library.wellcome.ac.uk and http://medphoto.wellcome.ac.uk).

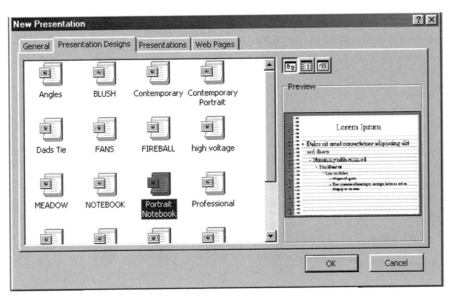

Figure 6.2 Choice of presentation templates

You can also use a search engine to find suitable photographs or images (or even video clips). Check out these resources for yourself and insert an image of a citrus fruit into the presentation. Whatever the source of your image, you will probably need to re-size it to fit the available space on your slide. Drag one of the control handles until the image is the correct size and placed correctly into position.

Building the presentation

You are now ready to move on to the next slide, so click on Insert: New Slide. This time choose the Bulleted list. Type in the information on Slide 2 in the presentation.

Although the presentation is not yet ready, you may want to preview what it looks like so far. Clicking on the Slide Sorter button will show the slide in your current set in outline view. To start the show, locate the Slide Show button and click on it. The first slide should now be displayed and will occupy the full screen. You may move backwards and forwards in a presentation by using the reverse and forward arrow keys, otherwise a single click of the mouse will move forward through the slide show, and, as we shall see later, activate the animated options.

If you move your mouse whilst viewing a slide a small button appears at the bottom left-hand corner. If you click on this button, a pop up menu appears that will provide some useful commands whilst looking at a slide show. Alternatively click with your right mouse button anywhere on the slide in view. Choose the Pen option from the menu and your mouse pointer becomes a pen. This enables you to be able to draw on your slide, by dragging the pen across the screen. Any drawing that is made is temporary and cannot be saved into your presentation. From the same menu, clicking on Pointer Options, allows you to change the colour of the lines that you draw. Try this out by clicking on an appropriate colour to contrast with your background. The feature is useful if you are explaining a complex diagram or want to emphasize a particular word or phrase on a slide.

While in Pen mode you cannot move on to the next slide by clicking the mouse.

Click again on the right-hand mouse button and select Arrow from the pop up menu to restore the features of the mouse pointer (or press N on the keyboard to move to the next slide). Each slide in the presentation can be viewed until you reach the end. To terminate the slide show without looking through them all, press the Esc key. If this does not return you to the slide view, click the Slide View button.

Notes Page View

This is a useful option if you want to prepare a set of speaker notes to go with your presentation. By selecting the Notes Page View button the screen splits into two parts: at the top is your slide and in a separate box below is a space in which to type your notes. Usually the notes frame is so small that you cannot clearly read any text entered into it, so use the Zoom control on the formatting toolbar and increase the zoom factor to 100 per cent. You should now be able to easily read any text that you enter into the notes page. The notes pages can then be printed out and used to refer to key points during the presentation or given as a handout to your audience.

Slide View

Click back on the Slide View button and display Slide 2. Now continue with your presentation by inserting a new slide. For Slide 3 we need to choose the Table slide. Enter the title and then double click, as shown on screen to add the table. You will be prompted to enter the number of rows and columns to complete the table; enter 2 columns and 5 rows and confirm your choice. The table will then appear in which you will be able to insert the information provided.

Inserting graphs and charts

Choose the option of inserting a new slide and this time select Chart slide (although you could also use Text and Chart/Chart and Text if you wanted to include some notes with the chart).

When the new slide is displayed you can double click on the Insert chart button to produce a datasheet and graph. You may enter the data directly onto the datasheet and the graph will be automatically plotted, or import data from a text file in Word, or from an Excel worksheet or insert a chart directly from Excel. It is usually more convenient to create a chart in Excel and then paste it into PowerPoint. Open Excel and insert the information given on the datasheet in Figure 6.3 and create the chart required for the slide. When you have completed the graph in Excel, copy and paste it into the space for the chart in PowerPoint. Re-size as appropriate and add the title.

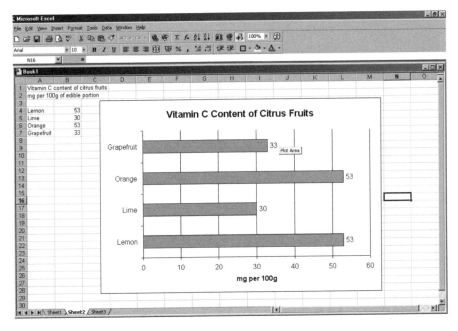

Figure 6.3   Preparing a graph using Excel

Drawing shapes on slides

You may create your own drawings or flow diagrams using the draw facility in PowerPoint. To use this option click on View Toolbars then ensure that the Drawing option is selected and click on OK. Tools may be used to:

- draw rectangles
- draw squares
- draw circles
- draw ellipses
- create shadows
- change line colour
- use autoshapes – choose from a selection including arrows, parallelograms, stars, etc.

Using the Draw features we will create a flow diagram for the biosynthetic pathway shown in Figure 6.4 to put into the presentation.

Firstly insert the names of the intermediates in the pathway by choosing the textbox option from the Draw toolbar and typing in the name of the substance. A border may be added to the textbox by clicking on the edge of the textbox and entering edit mode by clicking the right mouse button and choosing Format Text Box from the menu. The colour and style and weight (thickness) of the border may be selected.

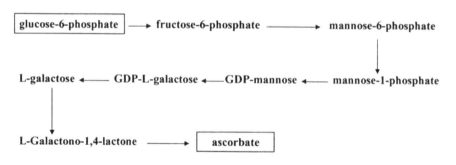

Figure 6.4   Example biosynthetic pathway – vitamin C (ascorbate) synthesis

Drawing lines/arrows

There are lines and arrows of different appearances available using the options from the Autoshapes on the Draw toolbar in addition to a plain line or arrow whose properties may be adjusted in the same way as the textbox border. The arrow tool is used by dragging it between its start and end points. By holding down the Shift key before drawing the line or arrow, you can control whether it is drawn vertical, horizontal or at a 45-degree angle.

Repeating this but using the Control instead of Shift key causes the line to be drawn from a central point.

Rotating an object

This may be achieved by selecting an object (e.g. an arrow) and then clicking on the Free Rotate tool located on the Draw toolbar. The mouse pointer changes to a rotating arrow. You are then able to rotate the position of an object until you are satisfied with its new position.

The tools in Draw are very easy to use; you can try experimenting with some of the other tools such as 3-D and shadow effects to create a three-dimensional appearance to some of the other objects. Once the object has been selected, click on the tool button and make a choice from the various formats offered. If you do not like the effect then simply Undo and try a different option.

Insert the information on Slide 6 from Figure 6.1 and then save your work to disk.

Adding animation effects

These are the effects for which PowerPoint has gained it reputation and certainly helps to enliven a presentation. The animation and sound effects should be used very carefully as they can become overwhelming for your audience unless they are used with restraint.

We can 'build' a slide so that bullet points appear as the presentation proceeds. There are various options on how you can bring in new points and what appears on screen with existing text that you have displayed. Complete the presentation by entering the information on the final slide shown in Figure 6.1.

Creating animation effects

Make certain that you are currently in Slide View and have Slide 2 displayed. If you click on the first sentence a frame will appear around all of the text. We will now set the animation features so that each bullet point on the slide appears separately instead of all at once as at present. From the Slide Show option select Custom Animation. A dialogue box now appears from which you will be able to customize the animation. From the Animation Settings click on the Effects tab. Firstly set the way in which the text will appear on the screen (currently set at No effect); a number of different styles are listed, select Fly From Left. On the right of this box under Introduce Text, click on All at once, then under Grouped By and select 1st Level Paragraphs. When text appears on the screen it will now come in from the left-hand side and one line will appear with each click of the mouse button. To add sound effects, change the option from No Sound to Camera (the third option will only be available if the computer you are working on is set up for sound). To determine what happens to the text once it has appeared, select Don't Dim. If you now move to the slide show you should find that the slide appears with its title. With the first click of the mouse button the first bullet point appears making the sound of a camera shutter closing as it does so; on the second click the first line disappears and the second bullet point flies in from the left. This is repeated for the remaining lines of text until the slide has been displayed and we move on to Slide 3 in the sequence. Press the Esc key to end the presentation.

Transition effects

Transition effects control the way in which slides appear on screen. Like text they can be animated in a variety of ways. Place Slide 3 in the current Slide View mode. From the Slide Show menu select Slide Transition. A number of different transitions are listed under the Effects, for which you are also able to change the speed from Slow to Medium or Fast. From the list of options choose a style and speed and watch the preview to see whether this is an effect that you like. Once you have made your selection, choose Apply All to apply the transition to all of the slides in the show. Leave the Advance section to Only on Mouse Click (so that you retain manual control over the presentation). Then return to the first slide in the sequence and run the show.

Sorting slides

The slides may be sorted using the Slide Sorter button shown at the bottom left-hand corner of the document window. Working in Slide Sorter you can select the slide timings for some or all of the slides in your presentation, without having to change views.

Click on Slide 2. You will see a black border appear to indicate that this slide has been selected. Hold down the Shift key and click on Slide 3.

Both Slides 2 and 3 should now be highlighted with the border. You may now alter the time that both of these slides appear on screen. This is useful for any unattended presentations, but can be used to keep to within set time limits for a live presentation, whilst allowing you to increase the time for more complex slides that need to remain on view for longer.

Click on Slide Show|Slide Transition to display the Transition dialogue box.

Under Advance, select the Automatically After [blank] seconds option button.

For each bullet point on a slide, allow at least 3 seconds, so enter 10 as the number of seconds that the slide will be displayed before advancing to the next slide. Click Apply to finish.

Click on Slide 1 and, using the instructions above, set its timing to 6 seconds.

If you click on Slide Show now, your timings will not take effect. Firstly you must click on Slide Show and use the pull down menu to display the Set Up Show dialogue box.

In the From: box type 1 then in the To: box type 3. In Advance, click on the Use Slide Timings box. Now click on the Show button and your timings will take effect.

Continuous loops

This allows the continuous display of a PowerPoint presentation until the Esc key is pressed.

Click on Slide Show and display the Set Up Show dialogue box.

In the From: box type 1 and in the To: box type six (as there are six slides in the show).

In Advance select the use Slide Timings option. Click on Loop Continuously Until 'Esc' option then click on the Show button. You should now see the show continuously loop round. When you have finished watching the presentation, press Esc to end the repeats.

Adding a slide

While you are in Slide View, click on Insert New Slide by Slide 2. You can then choose what type of slide to insert. Insert a bulleted text slide. This is just as easily deleted by re-selecting the slide and pressing the ⟨Del⟩ on the keyboard.

Ending a show

If you are not using a continuous loop it is a good idea to have a blank slide between the last slide and the first in the show. Without the blank slide, if you should click the mouse button accidentally whilst on the final slide PowerPoint will return to Slide View which would look unprofessional. This will not be necessary if you are running a continous loop.

Spelling

As in any item of work, you need to check your spelling. In addition to manual proofreading, you may activate the spellcheck facility by clicking on Tools then Spelling. The spellchecker works in the same way as in Word. Having checked that you now have all six slides in the presentation as shown in Figure 6.1, save your work and we will then create a handout.

Creating handouts

You are now ready to run your presentation and prepare the handouts for your audience. Click on View: Master then Handout Master. A handout sheet template now will appear for your presentation. You can insert information about your presentation in the header and footer. Insert your name in the top left of the handout. If you need to alter the layout of the handout (e.g. to landscape orientation as opposed to portrait, click on File: Page Setup and adjust it here, making sure that you select the option for A4 paper in this process). Now if you are ready to print, go to the File: Print option. Make sure that you select to print in Black and White if you don't want colour versions of your slides. Under Print What, select Handouts (six slides per page). Print the handout prepared from your presentation.

Tip: For some templates it is better to remove the background altogether (as in the example given) to print clear handouts. To do this, you will need to go back to the Format|Apply Design Template menu and select a Blank presentation. If one is not available on your system the choose Format|Background and in the pop up box tick the Omit Background Graphics from Master Check box and apply this to all the slides in the presentation (Apply to All).

6.3 Poster presentations

PowerPoint can be used equally as successfully to produce posters, large or small. You will need to choose the blank slide from the list of slide formats. Using File: Page Setup format the poster to the size you require; you may need to use the custom size option if the dimensions are not standard. Text is added to the slide using the textbox function and photographs and clipart or graphs and tables can be added in the same way as a slide presentation. In preparing a poster you should keep the following points in mind:

- Plan the poster carefully by making a rough design on a piece of paper before committing yourself to adding items on PowerPoint.

- Keep the design as simple as possible so that it doesn't look cluttered and disorganized. If it is a poster describing some research, it is useful to keep to the standard headings of Introduction, Methods, Results, Discussion and Conclusion.

- Maintain a consistent style throughout for your colour scheme and text. Make sure that writing is large enough to be read at a distance and that only key points are presented, there is not going to be enough room for detail.

- Use a plain background, if it is too busy it will detract from the other information on the poster. But make the display attractive by including illustrations and figures.

You will find some examples of posters on the book's support website together with PowerPoint presentations including animated features. Presenting the results of your research is the climax to a lot of hard work, so PowerPoint can be used with great effect to display your achievements to the rest of the scientific community. Have fun!

WEB SUPPORT – SECTION 6

This part of the website will show how PowerPoint can be used for slides and posters, showing a number of different formats and a list of further tips on making the most of your presentations.

Appendix

Further reading

An Introduction to Experimental Design and Statistics for Biology
David Heath (1995)
Routledge

Biosciences on the Internet
Georges Dussart (1992)
John Wiley & Sons

Experimental Design for the Life Sciences
Graeme D Ruxton & Nick Colegrave (2003)
Oxford University Press

Statistics with Applications to the Biological and Health Sciences
M. Anthony Schork & Richard D. Remington (2000)
Prentice Hall

Data Analysis and Presentation Skills by Jackie Willis.
© 2004 John Wiley & Sons, Ltd   ISBN 0470852739 (cased)   ISBN 0470852747 (paperback)

Alt key codes for special symbols in Microsoft applications

The following codes are for symbols that can be used in all Microsoft applications. To insert press the Alt key then the appropriate numeric code on the number key pad. On releasing the Alt key the symbol will be inserted.

Alt code	Symbol	Altcode	Symbol
129	ü	0131	f
130	é	0134	†
131	â	0137	‰
132	ä	0139	‹
133	à	0150	–
134	å	0151	—
135	ç	0153	TM
136	ê	0155	›
137	ë	0156	œ
138	è	0169	©
139	ï	0171	«
140	î	0174	®
141	ì	0175	–
145	æ	0176	°
146	Æ	0177	±
156	£	0178	²
159	f	0179	³
166	ª	0181	µ
167	º	0185	¹
171	½	0186	º
172	¼	0187	»
174	«	0188	¼
175	»	0189	½
225	ß	0190	¾
230	µ	0215	×
241	±	0247	÷
246	÷	0248	ø
248	°	0255	ÿ
253	²		

Statistical tables

Critical values for the Student *t*-test

df	Level of significance (*P*)						
	0.05	0.025	0.01	0.005	0.001	0.0005	One-tailed test
	0.1	0.05	0.02	0.01	0.002	0.001	Two-tailed test
1	6.314	12.706	31.821	63.657	318.31	636.62	
2	2.920	4.303	6.965	9.925	22.327	31.598	
3	2.353	3.182	4.541	5.841	10.214	12.924	
4	2.132	2.776	3.747	4.604	7.173	8.610	
5	2.015	2.571	3.365	4.032	5.893	6.869	
6	1.943	2.447	3.143	3.707	5.208	5.959	
7	1.895	2.365	2.998	3.499	4.785	5.408	
8	1.860	2.306	2.869	3.355	4.501	5.041	
9	1.833	2.262	2.821	3.250	4.297	4.781	
10	1.812	2.228	2.764	3.169	4.144	4.587	
11	1.796	2.201	2.718	3.106	4.025	4.437	
12	1.782	2.179	2.681	3.055	3.930	4.318	
13	1.771	2.160	2.650	3.012	3.852	4.221	
14	1.761	2.145	2.624	2.977	3.787	4.140	
15	1.753	2.131	2.602	2.947	3.733	4.073	
16	1.746	2.120	2.583	2.921	3.686	4.015	
17	1.740	2.110	2.567	2.898	3.646	3.965	
18	1.734	2.101	2.552	2.878	3.610	3.922	
19	1.729	2.093	2.539	2.861	3.579	3.883	
20	1.725	2.086	2.528	2.845	3.552	3.850	
21	1.721	2.080	2.518	2.831	3.527	3.819	
22	1.717	2.074	2.508	2.819	3.505	3.792	
23	1.714	2.069	2.500	2.807	3.485	3.767	
24	1.711	2.064	2.492	2.797	3.467	3.745	
25	1.708	2.060	2.485	2.787	3.450	3.725	
26	1.706	2.056	2.479	2.779	3.435	3.707	
27	1.703	2.052	2.473	2.771	3.421	3.690	
28	1.701	2.048	2.467	2.763	3.408	3.674	
29	1.699	2.045	2.462	2.756	3.396	3.659	
30	1.697	2.042	2.457	2.750	3.385	3.646	

Critical values of *T* for the Wilcoxon signed rank test

Level of significance (*P*)

	0.025 0.05	0.01 0.02	0.005 0.01	
	0.025	0.01	0.005	One-tailed test
	0.05	0.02	0.01	Two-tailed test
6	0	–	–	
7	2	0	–	
8	4	2	0	
9	6	3	2	
10	8	5	3	
11	11	7	5	
12	14	10	7	
13	17	13	10	
14	21	16	13	
15	25	20	16	
16	30	24	20	
17	35	28	23	
18	40	33	28	
19	46	38	32	
20	52	43	38	
21	59	49	43	
22	66	56	49	
23	73	62	55	
24	81	69	61	
25	89	77	68	
26	98	84	75	
27	107	92	83	
28	116	101	91	
29	126	110	100	
30	137	120	109	

Critical values for the Mann–Whitney *U*-test

Level of significance: 5% two-tailed test
2.5% one-tailed test

Size of largest sample (n_2)

Size of smallest sample (n_1)

	5	6	7	8	9	10	11	12	13	14	15	16	17	18	19	20	21	22	23	24	25	26	27	28	29	30
3	0	1	1	2	2	3	3	4	4	5	5	6	6	7	7	8	8	9	9	10	10	11	11	12	13	13
4	1	2	3	4	4	5	6	7	8	9	10	11	11	12	13	14	15	16	17	17	18	19	20	21	22	23
5	2	3	5	6	7	8	9	11	12	13	14	15	17	18	19	20	22	23	24	25	27	28	29	30	32	33
6		5	6	8	10	11	13	14	16	17	19	21	22	24	25	27	29	30	32	33	35	37	38	40	42	43
7			8	10	12	14	16	18	20	22	24	26	28	30	32	34	26	38	40	42	44	46	48	50	52	54
8				13	15	17	19	22	24	26	29	31	34	36	38	41	43	45	48	50	53	55	57	60	62	65
9					17	20	23	26	28	31	34	37	39	42	45	48	50	53	56	59	62	64	67	70	73	76
10						23	26	29	33	36	39	42	45	48	52	55	58	61	64	67	71	74	77	80	83	87
11							30	33	37	40	44	47	51	55	58	62	65	69	73	76	80	83	87	90	94	98
12								37	41	45	49	53	57	61	65	69	73	77	81	85	89	93	97	101	105	109
13									45	50	54	59	63	67	72	76	80	85	89	94	98	102	107	111	116	120
14										55	59	64	67	74	78	83	88	93	98	102	107	112	118	122	127	131
15											64	70	75	80	85	90	96	101	106	111	117	122	125	132	138	143
16												75	81	86	92	98	103	109	115	120	126	132	138	143	149	154
17													87	93	99	105	111	117	123	129	135	141	147	154	160	166
18														99	106	112	119	125	132	138	145	151	158	164	171	177
19															113	116	126	133	140	147	154	161	168	175	182	189
20																127	134	141	149	156	163	171	178	186	193	200
21																	142	150	157	165	173	181	188	196	204	212
22																		158	166	174	182	191	199	207	215	223
23																			175	183	192	200	209	218	226	235
24																				192	201	210	219	228	238	247
25																					211	220	230	239	249	258
26																						230	240	250	260	270
27																							250	261	271	282
28																								272	282	293
29																									294	305
30																										317

Index

Data Analysis and Presentation Skills by Jackie Willis.
© 2004 John Wiley & Sons, Ltd   ISBN 0470852739 (cased)   ISBN 0470852747 (paperback)